WORKED PROBLEMS IN OPTICS

By

Alan H. Tunnacliffe
BA, Dip. Maths, FBOA, DCLP

Senior Lecturer in Optics, Bradford College,
and Part-time Lecturer, University of Bradford.

Copyright May, 1979, A.H. Tunnacliffe

ISBN 0 900099 16 X

Published by The Association of Dispensing Opticians
22 Nottingham Place London W1M 4AT

Third Impression 1987

Printed in Great Britain by
The Eastern Press Limited
London & Reading

PREFACE

In ophthalmics, a large proportion of the examination questions in geometrical and physical optics papers consists of mathematically based problems. The author believes a student must spend considerable time in learning to solve such problems.

The effort involved is not inconsiderable. However, the reward gained is a deeper understanding of the theory involved and a facility for tackling problems.

There is a well established supply of worked problems books for students of mathematics and the sciences, but there is a dearth of this genre for opticians and optometrists. In general, text books do not have sufficient space for a great number of worked examples. The present text is to be regarded as a supplement to a main text.

After studying the theory the student should try the relevant questions from this book. If the problem can be completed then the answer confirms the student's grasp of the subject. If, however, the student cannot start or complete a problem, then the answer can be used to overcome the difficulty. The student should not treat these worked problems as examples simply to be read. He/she must struggle with the question and only use the answer when a hint or a confirmation is required.

The second time around, however, the answers may prove useful for revision.

There is a companion volume dealing with ophthalmic lens problems. Some questions in that text are given over entirely to qualitative answers.

Questions marked with an asterisk are of a slightly more difficult nature.

Bradford   Alan H. Tunnacliffe
May, 1979

THE GREEK ALPHABET

A number of symbols from the Greek alphabet are used in
the text. For convenience the whole alphabet is given
below, together with each letter's name.

A	α	alpha
B	β	beta
Γ	γ	gamma
Δ	δ	delta
E	ε	epsilon
Z	ζ	zeta
H	η	eta
Θ	θ	theta
I	ι	iota
K	κ	kappa
Λ	λ	lambda
M	μ	mu
N	ν	nu
Ξ	ξ	xi
O	ο	omicron
Π	π	pi
P	ρ	rho
Σ	σ	sigma
T	τ	tau
Y	υ	upsilon
Φ	φ	phi
X	χ	chi
Ψ	ψ	psi
Ω	ω	omega

CONTENTS

This 1987 reprint of the second edition has been reduced in size for students' convenience. Any scale diagram will be 80% of original size; i.e. where a scale is quoted as 1 cm $\equiv 1^{\Delta}$, the new scale will be 8 mm $\equiv 1^{\Delta}$.

1.   THE PROPAGATION OF LIGHT

1. The velocity of light in vacuo is 3×10^8 ms^{-1}. What will be the wavelengths corresponding to the following frequencies?

Orange 4.5×10^{14} Hz
Blue 6.96×10^{14} Hz.

Since $c = \nu\lambda$, where c is the speed of light in vacuo, ν is the frequency, and λ is the wavelength, we have on rearranging $\lambda = c/\nu$.

For orange light, $\lambda = c/\nu = 3 \times 10^8/4.5 \times 10^{14} = 667 \times 10^{-9}$m $= 667$nm.
For blue light, $\lambda = c/\nu = 3 \times 10^8/6.96 \times 10^{14} = 431$ nm.

---

2.   A cone of light diverges from a point source through a rectangular aperture 3cm × 4cm distant 20cm from the source. Find the area of the patch of light on a screen parallel to and 100cm from the aperture.

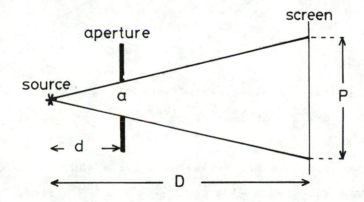

Fig. 1.1

Let P be one dimension of the patch of light and a be the corresponding dimension of the aperture. The distances d and D are as shown in figure 1.1.

Then P=aD/d , from similar triangles. Hence, the length of the patch is $4 \times 120/20 = 24$cm, and the breadth is $3 \times 120/20 = 18$cm. The area is length × breadth = $24 \times 18 = 432$cm^2.

---

3. A spherical source of light 10cm in diameter is placed 50cm
    from a circular aperture of 16cm diameter in an opaque screen,
    figure 1.2. Find the size and nature of the patch of light
    on a white screen 150cm from and parallel to the plane of the
    aperture.

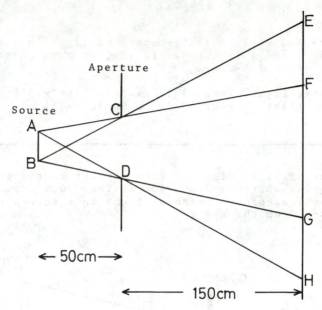

Fig. 1.2

(i) Using similar triangles ABC and CEF,

$$EF = CF.\frac{AB}{AC} = AB.\frac{CF}{AC} = AB \times \frac{150}{50} = 10 \times \frac{150}{50} = 30\text{cm},$$

since CF:AC = 150:50 .

(ii) Now, using similar triangles BCD and BEG,

$$EG = BG.\frac{CD}{BD} = CD.\frac{BG}{BD} = CD \times \frac{200}{50} = 16 \times \frac{200}{50} = 64\text{cm},$$

since BG:BD = 200:50 .

Hence, FG = EG - EF = 64 - 30 = 34cm, and GH = EF = 30cm.

The diameter of the fully illuminated area is FG = 34cm,
and the diameter of the whole patch is EH = EF+FG+GH = 94cm.

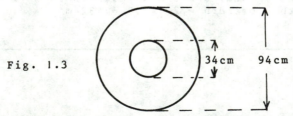

Fig. 1.3

4.   In a pinhole camera the distance between the object and the image is 5m. Find the length of the camera if the image is one-fiftieth the size of the object, and the distance of the object from the pinhole.

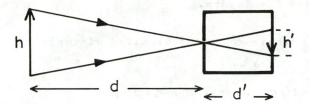

Fig. 1.4
Let d' be the length of the camera. Then d+d' = 5 ....(1)
and

$$\frac{h'}{h} = \frac{d'}{d} = 0.02 \qquad \ldots\ldots\ldots\ldots\ldots\ldots\ldots(2)$$

$$\text{or } d' = 0.02d$$

Substituting for d' in equation (2) into equation (1),

$$d+d' = d+0.02d = 5$$

Hence,   d = 5/1.02 = 4.90m. This is the distance of the object from the pinhole. From equation (1), the length of the camera is

$$d' = 5-d = 5-4.9 = 0.1m.$$

---

5.   A pinhole camera produces an image 2.25cm long, and when the screen is withdrawn 3cm further from the pinhole the length of the image increases to 2.75cm. What was the original distance between the pinhole and the screen?

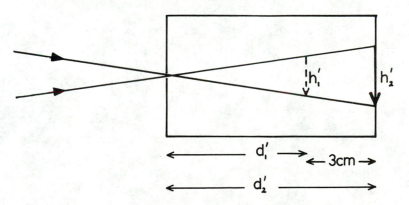

Fig. 1.5

The angular subtent of the image at the pinhole remains constant. From the figure,

$$\frac{d_1'}{d_2'} = \frac{h_1'}{h_2'} \quad \text{where } d_1' \text{ is the required distance.}$$

Rearranging and noting that $d_2' = d_1' + 3$,

$$d_1' = d_2' \cdot h_1'/h_2' = (d_1' + 3)h_1'/h_2' \ ,$$

whence, $\quad d_1' = 3\frac{h_1'}{h_2'} / (1 - \frac{h_1'}{h_2'}) = 3 \times \frac{2.25}{2.75} / (1 - \frac{2.25}{2.75}) = 13.5\text{cm}.$

-5-

2. REFRACTION AND REFLECTION - GENERAL PROBLEMS

1.   The velocity of light in vacuo is 3×10^8 ms^{-1}. Find the velocity in glass of refractive index 1.5.

   By definition, $\dfrac{\text{velocity in vacuo}}{\text{velocity in glass}} = \text{refractive index}.$

Hence, velocity in glass $= \dfrac{\text{velocity in vacuo}}{\text{refractive index}} = \dfrac{3 \times 10^8}{1.5} = 2 \times 10^8$ ms^{-1}

Note: refractive index implies a specific frequency of visible light.

---

2.   Calculate the angle of refraction for a ray incident in air at 30^0 on a block of crown glass (n_g = 1.52).

   Using Snell's Law, $n \sin i = n' \sin i'$,

   $\sin i' = \dfrac{n}{n'} \sin i = \dfrac{1}{1.52} \sin 30^0 = 0.3289.$

   Whence, $i' = \arcsin 0.3289 = 19.2^0$

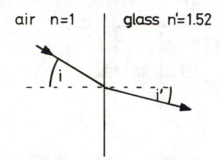

air n=1   glass n'=1.52

Fig. 2.1

---

3.   The absolute refractive indices of water, glass, and oil are 1.33, 1.65, and 1.45, respectively. Calculate the relative refractive indices for light being refracted from
         a) water into glass,
         b) oil into water,
         c) glass into water.

   a) Let n_w and n_g represent the absolute refractive indices of the water and glass, respectively. Then, the relative refractive index is

   $_w n_g = \dfrac{n_g}{n_w} = \dfrac{1.65}{1.33} = 1.24$.

   b) In similar vein,

   $_o n_w = \dfrac{n_w}{n_o} = \dfrac{1.33}{1.45} = 0.917$

c)
$$_g n_w = \frac{n_w}{n_g} = \frac{1.33}{1.65} = 0.806 \quad .$$

Note that $_g n_w = \frac{1}{_w n_g}$.

4.   A flag has three vertical stripes which appear red, white, and yellow in white light. Describe their appearances under illumination with a) red light and b) green light. Further, if the flag is illuminated with white light describe its appearance when viewed through c) a yellow glass and d) a red glass.

Figure 2.2   shows the flag seen under white light

Fig. 2.2

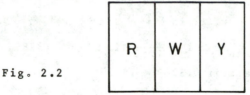

The answers are presented in the following diagram.

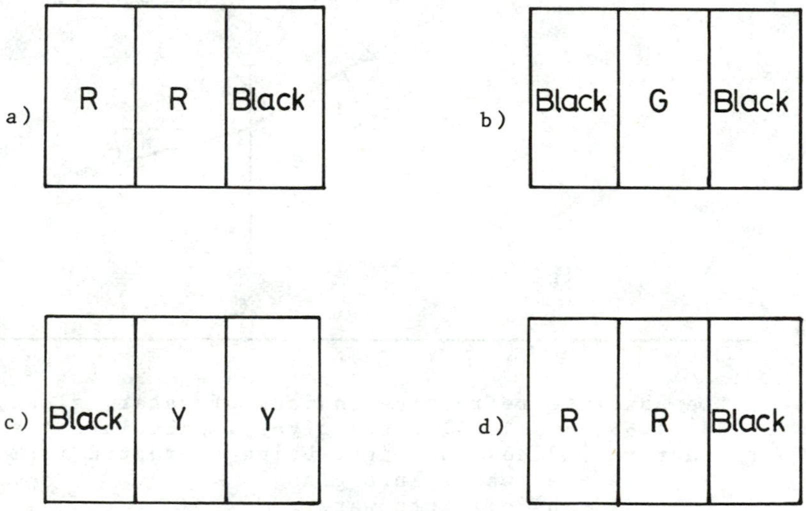

a)

b)

c)

Fig. 2.3

d)

Notes: (i) We are assuming a red strip reflects only a single wavelength from the red part of the visible spectrum, all other wavelengths being absorbed.
        (ii) A red glass is assumed to transmit only one wave-
             length which is identical to the wavelength in (i).
       (iii) Both assumptions are idealisations
        (iv) Similar assumptions apply where yellow is involved.

3. REFLECTION: PLANE SURFACES

1.   A sight-testing chart, measuring 120cm by 80cm with its longer dimension vertical, is to be viewed monocularly by reflection in a plane mirror. If the observer's eye and the chart are each 3m from the mirror, find the minimum size of mirror that can be used to see the whole chart.

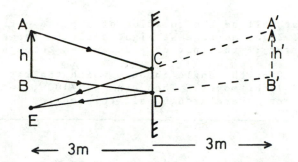

Fig. 3.1   ←— 3m ————   —— 3m →

A'B', the image of the chart, is 3m behind the mirror. Triangles ECD and EA'B' are similar. Since EB' = 2ED, h' = 2CD. But h' = h, so CD = $\frac{1}{2}$h' = $\frac{1}{2}$h. Hence, the mirror must be 60cm × 40cm with its longer axis vertical.

---

2.   In the previous question if the observer's eye is 1.3m above the floor, and the lower edge of the mirror is 1.5m above the floor, how high is the lower edge of the chart?

Draw a horizontal through E in figure 3.1. This is illustrated in figure 3.2 together with the line EDB'.

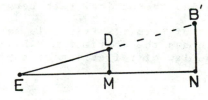

Fig. 3.2

Drop perpendiculars DM and B'N to the horizontal line EMN. From the similar triangles EDM and EB'N ,

$$B'N = \frac{EN}{EM} \cdot DM = \frac{6}{3} \cdot DM = 2DM .$$

Now, E is 1.3m above the floor and D is 0.2m higher; i.e. DM = 0.2m. So, B'N = 0.4m and the lower edge of the chart is 1.3 + 0.4 = 1.7m above the floor.

---

3.   Stand in front of a mirror and use a chinagraph pencil to draw round the outline of your head. Are you surprised at the size of the outline? If so, refer to question 3.1 .

4.   What will be the rotation of a plane mirror reflecting a spot of light on to a straight scale (a tangent scale) 1m from the mirror if the spot of light moves through 5cm?

Let θ be the angle through which the ray to the centre of the spot of light moves. Then $\tan \theta = 5/100$, so $\theta = \arctan 0.05 = 2.86$ The mirror moves through half this angle.

5.*   Suppose a ray in going from B to C traverses distances ℓ_1, ℓ_2, ..., ℓ_m in media of indices $n_1, n_2, ..., n_m$, respectively. The total time taken is

$$t = \frac{\ell_1}{v_1} + \frac{\ell_2}{v_2} + \cdots\cdots + \frac{\ell_m}{v_m} = \sum_1^m \frac{\ell_r}{v_r}$$

But $v_r = \frac{c}{n_r}$, where v_r is the speed of light in the medium r and c is the speed of light in vacuo.

Hence, $t = \sum_1^m \frac{\ell_r}{v_r} = \frac{1}{c} \sum_1^m n_r \ell_r$.

$\sum_1^m n_r \ell_r$ is called the Optical Path Length, or O.P.L.

One statement of Fermat's Principle says that a ray traverses a route which has the shortest O.P.L.. Use Fermat's Principle to derive the law of reflection at a plane mirror.

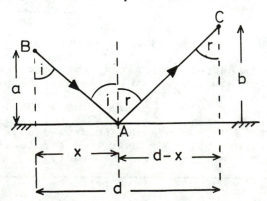

Fig. 3.3

* Question involving calculus.

If the medium above the mirror has an index n then the
$O.P.L. = n.BA + n.AC = n\sqrt{(a^2 + x^2)} + n\sqrt{(b^2 + (d-x)^2)}$.
The O.P.L. is a function of x and the light will take a path
for which

$$\frac{d(O.P.L.)}{dx} = 0$$

i.e. $nx(a^2 + x^2)^{-\frac{1}{2}} - n(d - x)(b^2 + (d - x)^2)^{-\frac{1}{2}} = 0$.

But, $x/(a^2 + x^2)^{\frac{1}{2}} = \sin i$ and $(d - x)/(b^2 + (d - x)^2)^{\frac{1}{2}} = \sin r$.

Hence, $n \sin i - n \sin r = 0$, from which $i = r$.

---

6. An erect pin is 8cm from a plane mirror. If the pin is
    moved 4cm towards the mirror how far does its image move?
    Alternatively, if the mirror is moved 4cm towards the pin
    again find the distance the image moves.

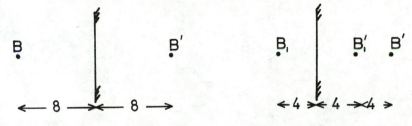

a) The initial situation

b) After the pin is moved
   through 4cm

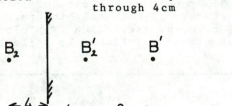

c) The mirror moved 4cm

Fig. 3.4 (Distances in cm)

Figure 3.4a shows the initial situation. In moving the
pin 4cm to position B_1 , figure 3.4b, the image moves 4cm
to position B_1'.

If, now, the pin remains stationary at B_2 the same
position as B_1, but the mirror is brought up to 4cm from the
pin, then the image is at B_2'. Clearly, $B_2'B'$ is 8cm.

7. Two mirrors are inclined at 90^0. Calculate the number of images produced and show their position on a diagram for an object placed between them.

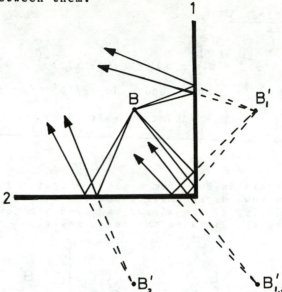

Fig. 3.5

If α is the angle between the mirrors the number of images formed is given by $360^0/\alpha - 1$, when α divides into 360^0 an integral number of times. In this case the number of images is 3. Images B_1' and B_2' arise by reflection from mirrors 1 and 2, and $B_{1,2}'$ by reflection at mirror 1 followed by reflection at mirror 2.

4. REFRACTION AT PLANE SURFACES

1. A beam of light with a circular cross-section, diameter 2cm, strikes a plane surface making an angle of 30^0 with the normal. Find the deviation of the beam on entering the second medium and its cross-sectional diameter in the plane of incidence. Assume the first medium is air and the second is glass of refractive index 1.5 .

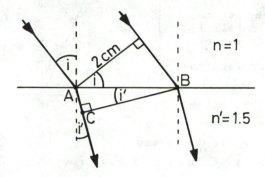

Fig. 4.1

Using Snell's Law, $n \sin i = n' \sin i'$, whence

$$\sin i' = \frac{n}{n'} \sin i = \frac{1}{1.5} \sin 30^0 = \frac{1}{1.5} \times \frac{1}{2} = \frac{1}{3} \ .$$

So, $i' = \arcsin(\tfrac{1}{3}) = 19.47^0$.

The deviation is $i - i' = 30^0 - 19.47^0 = 10.53^0$.
From the figure, the diameter of the beam in the plane of incidence in the glass is $BC = AB \cos i' = \frac{2}{\cos i} \cdot \cos i'$

$$= \frac{2}{\sqrt{3}/2}(0.9428) = 2.18 \text{cm}$$

(since $AB = 2/\cos i$).

---

2. A ray of light is incident on one face of a plane parallel sided glass plate of thickness t. Find an expression for the displacement of the emergent ray.

The displacement is
$CD = AC \sin \alpha$, where $\alpha = i - i'$.
But $AC = AB/\cos i'$, so

$$CD = AB \cdot \frac{\sin \alpha}{\cos i'} = t \frac{\sin(i - i')}{\cos i'},$$

since $AB = t$.

Fig. 4.2

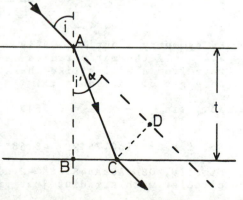

3. A collimated beam strikes a parallel sided glass block.
    Part of the beam reflects off the top surface and part off
    the bottom, figure 4.3 . Show that the two beams going back
    into the air are parallel.

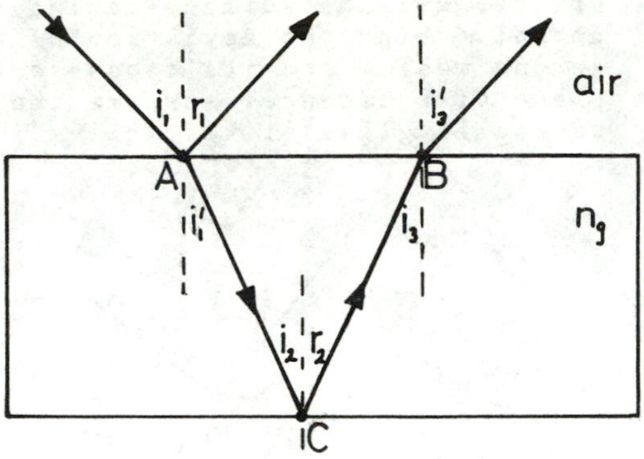

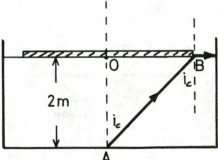

By the law of reflection
$i_1 = r_1$. It remains to
show that $i_3' = r_1$. Since
the normals at the points
A, B, and C are parallel,
$i_1' = i_2$ (alternate angles).

$i_2 = r_2$ (law of reflection),

and $r_2 = i_3$ (alt. angles).

So, $i_3 = i_1'$. At B, using

Snell's Law, $n_g \sin i_3 = \sin i_3'$.

But at A, $\sin i_1 = n_g \sin i_1' = n_g \sin i_3$.

Hence, $i_1 = i_3'$, and since $i_1 = r_1$, we have $r_1 = i_3'$.

Fig. 4.3

---

4. A point source of light lies at the bottom of a pool of
    water 2m deep. A thin cork mat floats on the water with
    its centre immediately above the source. Find the shape
    and the minimum size of the mat such that an observer above
    the surface cannot see the light. The refractive index
    of the water is 1.33 .

Fig. 4.4

By symmetry about AO, OB is the radius of a circle, figure
4.4 . If the mat's radius is just greater than OB all the
rays from A will strike the surface of the water making
an angle of incidence $i > i_c$.

Now, $\sin i_c = \dfrac{1}{n_w} = \dfrac{1}{1.33} = 0.7519$.

So, $i_c = \arcsin 0.7519 = 48.68^0$.

Finally, $OB = AO \tan 48.68^0 = 2 \times 1.137 = 2.27m$. The mat is
circular with a radius just greater than 2.27m.

5. ABCD are corners on one face of a glass cube. One face
   of the cube, with edge AB, is covered with white absorbent
   paper and is brightly illuminated; a dark screen covers
   face DA. E is a point on the face BC. When the paper is
   moistened with a liquid an observer looking into face BC
   along a direction FE sees a boundary between light and dark
   areas. Explain this. When the liquid on face AB is water,
   index 1.33, the angle FEC is $37^0 23'$; when the liquid is
   glycerol angle FEC is $45^0 15'$. Calculate the refractive
   indices of the glass and the glycerol.

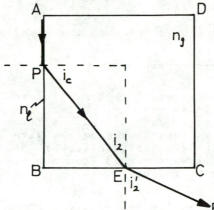

Fig. 4.5

Let n_l be the refractive index of the liquid and n_g the
index of the glass. i_c is the critical angle for the
liquid-glass boundary. Clearly, no light can reach E from
the face AB between the points A and P since the angle of
refraction would exceed i_c. We can find the refractive
index of the glass by considering Snell's Law at the points
P and E. Thus, $1.33 \sin 90^0 = n_g \sin i_c$

$$\text{or, } 1.33 = n_g \sin i_c \quad \dots\dots\dots\dots\dots\dots(1)$$

$$\text{Also, } n_g \sin i_2 = 1.\sin i_2' \quad \dots\dots\dots\dots\dots\dots(2)$$

Now $i_2' = 90^0 - \hat{FEC} = 90^0 - 37^0 23' = 52^0 37'$, and $i_2 = 90^0 - i_c$.

Substituting these values in equation (2),
$$n_g \sin(90^0 - i_c) = \sin 52^0 37' = 0.7946 \quad \dots\dots(3)$$

Using the fact that $\sin(90^0 - i_c) = \cos i_c$, we can rewrite(3),
$$n_g \cos i_c = 0.7946 \quad \dots\dots\dots\dots\dots\dots\dots\dots(4)$$

We can now solve (1) and (4) for n_g. Collecting them
together we have
$$n_g \sin i_c = 1.33$$

$$n_g \cos i_c = 0.7946 .$$

Squaring both sides and adding, and using $\sin^2 \theta + \cos^2 \theta = 1$
$$n_g^2 (\sin^2 i_c + \cos^2 i_c) = 1.33^2 + 0.7946^2 = 2.4 .$$

$$\text{So, } n_g = \sqrt{2.4} = 1.56 .$$

Knowing n_g we find the index of the glycerol as follows.

At E, Snell's Law gives $n_g \sin i_2 = 1. \sin i_2' = \sin(90^\circ - 45^\circ 15')$.

So, $\sin i_2 = \frac{1}{1.56} \cdot \sin 44^\circ 45' = \frac{1}{1.56} \times 0.7040 = 0.4513$.

whence $i_2 = 26^\circ 50'$.

Now, $i_c = 90^\circ - i_2 = 90^\circ - 26^\circ 50' = 63^\circ 10'$, so refraction at P gives

$$n_\ell \sin 90^\circ = n_g \sin i_c$$

$$n_\ell = 1.56 \times \sin 63^\circ 10' = 1.56 \times 0.8923 = 1.39 .$$

That is, the refractive index of the glycerol is 1.39.

---

6. The base of a glass vessel is 3cm thick and is made of glass with refractive index $n_g = 1.5$. Water, $n_w = 1.33$, is poured into the vessel to a depth of 4cm. A layer of paraffin oil of depth 4.32cm is then poured on the water. A mark on the lower surface of the glass base appears 8cm below the upper surface of the oil when viewed from above. What is the refractive index of the oil?

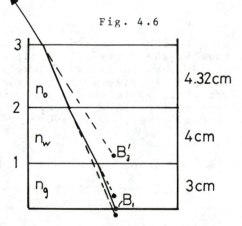

Fig. 4.6

Since each of the surfaces 1, 2, and 3 are zero power, then L' = L for each surface. Now for the first surface
$L = \frac{n}{\ell} = \frac{1.5}{-0.03} = -50D$.

So $L' = -50D$ and $\frac{n'}{\ell'} = L'$

whence $\frac{\ell'}{n'} = \frac{1}{-50} = -0.02m$.

Arriving at the second surface the light appears to have travelled 2cm plus the air equivalent thickness between surfaces 1 and 2. That is, $2cm + 4/1.33cm$, which is 5.00cm. A similar statement holds for the third surface where the light reaching it has travelled an air equivalent distance of $5.00 + 4.32/n_o$ cm, where n_o is the refractive index of the oil.
All of this is equal to the air equivalent distance of 8cm.
Thus
$$5.00 + \frac{4.32}{n_o} = 8 \text{ whence } n_o = 1.44 .$$

Knowing the trick now we can right away sum the air equivalent distances and equate to 8cm.
$$\text{Thus, } \frac{3}{1.5} + \frac{4}{1.33} + \frac{4.32}{n_o} = 8 \text{ from which the same}$$
result follows.

---

7. A ray of light passes through a glass prism. In what
    circumstances is the deviation a minimum?
    An equiangular prism ABC transmits sodium light, the light
    entering the face AB and leaving through the face AC. When
    the deviation is a minimum the angle of incidence is 58.63^0.
    Calculate the refractive index of the glass and the angle of
    minimum deviation.

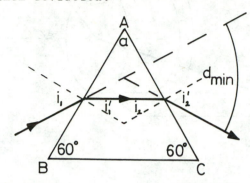

Fig. 4.7

Minimum deviation occurs when $i_1 = i_2'$. As a result $i_1' = i_2$
and , since $i_1' + i_2 = a$, $i_1' = i_2 = a/2$.
The deviation is given by $d = i_1 + i_2' - a$, and when d is a
minimum $i_1 = i_2'$ so,

$$d_{min} = 2i_1 - a = (2 \times 58.63^0) - 60^0 = 57.26^0.$$

The refractive index of the glass is given by

$$n = \frac{\sin(\frac{a+d_{min}}{2})}{\sin\frac{a}{2}} = \frac{\sin(\frac{60^0+57.26^0}{2})}{\sin\frac{60^0}{2}} = 1.707,$$

or, alternatively, note that $i_1' = a/2 = 30^0$ in minimum
deviation. Hence, since $i_1 = 58.63^0$,

$$n = \frac{\sin i_1}{\sin i_1'} = \frac{\sin 58.63^0}{\sin 30^0} = 1.707 .$$

---

8. Figure 4.8 shows a thin glass fibre (n_f) surrounded by a
    lower index cladding (n_c). There is a maximum incident
    angle i_{max} such that any ray striking the face at $i > i_{max}$
    will arrive at an internal wall at an angle less than i_c
    and will not be totally internally reflected.
    Show that

$$\sin i_{max} = \frac{1}{n_0}(n_f^2 - n_c^2)^{\frac{1}{2}} .$$

(Note: as long as $i << i_{max}$ light will be repeatedly internally
reflected down the length of the cylinder: this is the
basis of fibre optics).

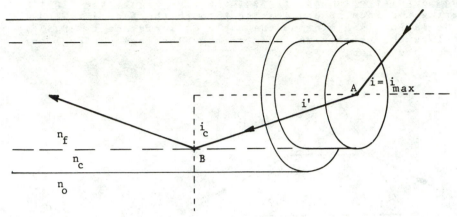

Fig. 4.8

From the figure, if $i > i_{max}$ the ray striking the inner surface of the cylinder will make an angle with the normal at B less than i_c. It will then pass into the cladding and will not be repeatedly internally reflected along the fibre. Hence, i_{max} occurs when the angle of incidence at B is i_c.

Refraction at B providing a 'grazing ray' emerging from the fibre into the cladding gives, by Snell's Law,

$$n_c \sin 90^0 = n_f \sin i_c, \quad \text{or} \quad \sin i_c = n_c/n_f \quad \ldots\ldots\ldots\ldots(1)$$

Now, $i' = 90^0 - i_c$, so at A,

$$n_o \sin i_{max} = n_f \sin i' = n_f \sin(90^0 - i_c) = n_f \cos i_c \quad ..(2)$$

Rearranging equation (2),

$$\sin i_{max} = \frac{n_f}{n_o} \cos i_c = \frac{n_f}{n_o}(1 - \sin^2 i_c)^{\frac{1}{2}}$$

$$= \frac{n_f}{n_o}\left(1 - \frac{n_c^2}{n_f^2}\right)^{\frac{1}{2}}, \text{ using equation (1).}$$

$$= \frac{n_f}{n_o}\left(\frac{n_f^2 - n_c^2}{n_f^2}\right)^{\frac{1}{2}}$$

$$= \frac{1}{n_o}(n_f^2 - n_c^2)^{\frac{1}{2}}.$$

N.B. Use has been made of the relationships $\sin(90-x) = \cos x$ and $\sin^2 x + \cos^2 x = 1$, or $\cos x = (1 - \sin^2 x)^{\frac{1}{2}}$.

9. Show that if a prism of refractive index n is to give
minimum deviation d, its refracting angle a is given by

$$\tan\frac{a}{2} = \frac{\sin\frac{d}{2}}{n-\cos\frac{d}{2}} \quad .$$

The usual equation relating a, d, and n is

$$n = \frac{\sin(\frac{a+d}{2})}{\sin\frac{a}{2}} \quad .$$

Expanding the numerator of the R.H.S. using the identity
sin(A+B) = sinAcosB + cosAsinB, we have

$$n = \frac{\sin\frac{a}{2}\cos\frac{d}{2} + \cos\frac{a}{2}\sin\frac{d}{2}}{\sin\frac{a}{2}}$$

$$= \cos\frac{d}{2} + \cot\frac{a}{2}\sin\frac{d}{2} \quad , \text{ since } \frac{\cos x}{\sin x} = \cot x.$$

Rearranging,

$$\cot\frac{a}{2} = \frac{n - \cos\frac{d}{2}}{\sin\frac{d}{2}}$$

and taking reciprocals

$$\tan\frac{a}{2} = \frac{\sin\frac{d}{2}}{n - \cos\frac{d}{2}} \quad .$$

Use has been made in the last step of the facts that

$$\frac{1}{\cot x} = \tan x \text{ and } \frac{1}{a/b} = \frac{b}{a} \quad .$$

---

10. An equilateral prism ABC is made of glass of refractive
index n_D = 1.5 . A narrow parallel beam of sodium light
is incident on the face AB. Find the angle of incidence
such that the beam will just be totally internally reflected
at the face AC. If the beam were white light explain what
happens at the face AC.

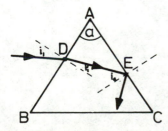

Fig. 4.9

A ray must make an angle of incidence at E just greater than

i_c in order to be internally reflected. Now, $\sin i_c = \dfrac{1}{n_D} = \dfrac{1}{1.5}$.

Hence, $i_c = \arcsin\dfrac{1}{1.5} = 41.81^0$.

In general, $i_1' + i_2 = a$,

or, $i_1' = a - i_2$.

Here $i_2 = i_c$, so $i_1' = a - i_c = 60^0 - 41.81^0 = 18.19^0$.

Snell's Law applied at the point D gives
$$\sin i_1 = n_D \sin i_1' = 1.5\sin 18.19^0 = 0.4683,$$
whence, $\qquad i_1 = \arcsin 0.4683 = 27.92^0$.

When the incident beam is white light wavelengths longer than those for sodium light will refract through the surface AC since the refractive index will be less than 1.5 . This results in a critical angle greater than 41.81^0, the critical angle for the sodium light. Hence, i_2 will be less than i_c for all these longer wavelengths and such rays will not suffer total internal reflection.
The converse is true for wavelengths shorter than those for sodium light. All such shorter wavelengths are totally internally reflected.

Note: a sodium source emits two wavelengths, very nearly the same, known as a doublet after the fact that they produce two images, very close together, of the collimator slit in the spectrometer. These two Fraunhofer lines, D_1 and D_2, are produced by the wavelengths 589.5923nm and 588.9953nm. When a refractive index is quoted as n_D the D is taken as the mean wavelength of the D_1 and D_2 lines, i.e. 589.2938nm.

5. <u>CURVATURE AND REFRACTION AT A CURVED SURFACE</u>

1.* Show that the curvature of a circle is 1/r, where r is
   the radius of the circle.

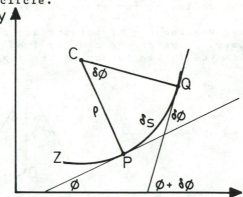

Fig. 5.1

Consider figure 5.1 . P and Q are 'very close together'
on the curve Z. The tangents at P and Q make angles ϕ and
$\phi + \delta\phi$ with the x-axis. δs is the length of the arc PQ
and ρ the length of the line CP which is perpendicular to
the tangent at P. The curvature of the arc PQ is defined as
the rate at which ϕ changes with respect to the arc length
PQ. Since ϕ increases by $\delta\phi$ in going from P to Q, a distance
δs around the arc, the definition can be written

$$\text{curvature} = \frac{\delta\phi}{\delta s}$$

To see the reasonableness of the definition imagine a steeper
curve where ϕ changes by 2.$\delta\phi$ in going δs around the arc,
then the curvature will clearly be twice as much.
Now, for small δs ,

$$\delta\phi = \frac{\delta s}{\rho}$$

$$\text{and the curvature} = \frac{\delta\phi}{\delta s} = \frac{\delta s/\rho}{\delta s} = \frac{1}{\rho} \ ,$$

where ρ is called the radius of curvature at the point P.
In a circle the radius of curvature is constant and is
denoted by r. Thus the curvature is 1/r and is the same at
all points on the circumference. This is illustrated in
figure 5.2.

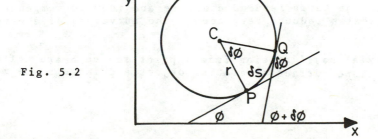

Fig. 5.2

2. The fixed legs of a spherometer form an equilateral triangle of side 3.41cm. When the instrument is placed on an optical flat the reading on the scale is zero. When it is placed on the convex surface of a lens the reading is 0.0971cm. Calculate the radius of curvature of the surface.

The reading gives the sag of the surface above the plane of the three feet. However, we need to find the distance from the central moveable leg to one of the fixed legs, y in figure 5.3 .

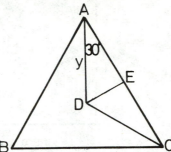

Fig. 5.3

AE = ½ × 3.41 = 1.705cm. In triangle ADE,

$$y = AE/\cos 30^0 = 1.705/\frac{\sqrt{3}}{2} = 1.969\text{cm}.$$

Using the approximate formula,

$$r = y^2/2s = 1.969^2/(2 \times 0.0971) \quad \text{cm}^2/\text{cm}$$
$$= 19.96\text{cm}.$$

Using the exact formula,

$$r = \frac{y^2}{2s} + \frac{s}{2} = \frac{3.877}{2 \times 0.0971} + \frac{0.0971}{2} = 20.01\text{cm}$$

Incidentally, the curvature is given by $R = \frac{1}{r}$ but curvature is given in units m^{-1} (or dioptre) which requires r in metres.

Hence, $R = \frac{1}{r} = \frac{1}{0.2001} = 4.998\text{m}^{-1}$ (or D) $\simeq$ 5 D.

___

3. Define the terms reduced distance and reduced vergence. Show that a reduced vergence is the curvature of a wavefront.

'Paraxial rays' arising from a point source B are refracted at a plane surface separating object and image space media

with refractive indices n and 1, respectively, figure 5.4 .
By the real and apparent
depth relationship $\ell' = \ell/n$.
ℓ' is the distance the emergent
light appears to have
travelled in air (index 1)
and is therefore called
the air-equivalent distance
which the light has traversed.
But $\ell' = \ell/n$, where ℓ is the
true distance the light has
travelled in the medium of
index n, and since n>1 the
quantity ℓ/n is called a
reduced distance. Hence,
since $\ell' = \ell/n$, the terms
air-equivalent distance and
reduced distance are
synonymous.

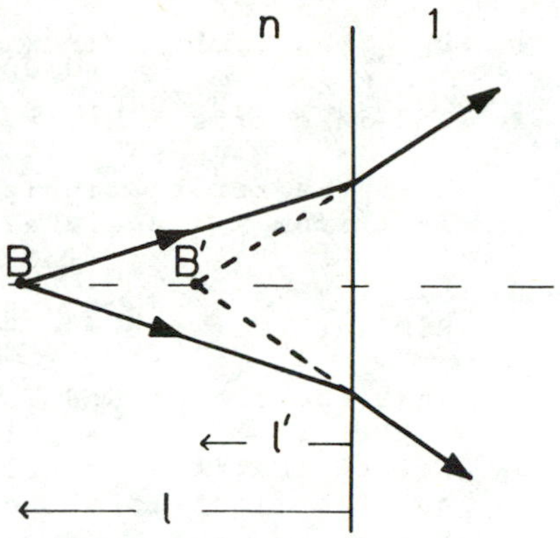

Fig. 5.4

If ℓ is measured in metres $\frac{1}{\ell/n}$ is a curvature. By analogy
with $R = \frac{1}{r}$, we write $L = \frac{1}{\ell/n}$, or more usually, $L = \frac{n}{\ell}$, and
this represents the curvature of the wavefront arriving at
the surface in figure 5.4. Because it is the reciprocal of
a reduced distance the quantity L is a 'reduced curvature'
which is given the special name REDUCED VERGENCE. The latter
term is often abbreviated to vergence although, strictly,
vergence ought to mean the curvature of a wavefront in air
when $\ell/n = \ell/1 = \ell$; i.e. it isn't reduced!

---

4. If the radius of curvature of a convex spherical surface is
6.00mm and the object and image space refractive indices are
1 and 1.3333, respectively, find
        (a) the positions of the first and second focal
            points,
        (b) the image position for an object 0.30m in
            front of the surface,
  and (c) the size of the image of an object subtending
            an angle of 2^0 at the vertex of the surface.

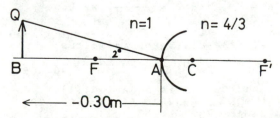

Fig. 5.5 (not to scale)

(a) The power of the surface is given by $F = \dfrac{n' - n}{r}$ and the first and second focal lengths are

$$f = -\frac{n}{F} \quad \text{and} \quad f' = \frac{n'}{F} \text{ , respectively.}$$

Now, $F = \dfrac{n' - n}{r} = \dfrac{1.3333 - 1}{+0.006} = +55.55D$; r is positive

and in metres. Hence, $f = -\dfrac{n}{F} = -\dfrac{1}{55.55} = -0.0180m \equiv -18.0mm;$

i.e. the first principal focal point, measured to the left from A since the length is negative, is

$$f = AF = -18.0mm.$$

Similarly, $f' = \dfrac{n'}{F} = \dfrac{1.3333}{55.55} = +0.0240m \equiv +24.0mm.$

That is, the second focal length AF', measured to the right from A since f' is positive, is 24.0mm.
Notice that the symbol F is used for both surface power and the first focal point. Some texts use P for power.

(b) (i) The image position can be obtained using the

Gaussian equation $\dfrac{n'}{\ell'} - \dfrac{n}{\ell} = \dfrac{n' - n}{r}$.

$r, \ell,$ and ℓ' must be in the same units.
The data are:
$$n = 1, \quad n' = 4/3, \quad \ell = -300mm, \quad r = +6mm.$$
We require ℓ': inserting the values into the equation

$$\frac{4/3}{\ell'} - \frac{1}{-300} = \frac{4/3 - 1}{6}$$

Rearranging, $\dfrac{4/3}{\ell'} = \dfrac{4/3 - 1}{6} - \dfrac{1}{300} = \dfrac{50/3 - 1}{300}$.

Taking reciprocals, $\dfrac{\ell'}{4/3} = \dfrac{300}{47/3}$

or, $\ell' = \dfrac{4}{3} \times \dfrac{300}{47/3} = \dfrac{4 \times 300}{47} = +25.5mm.$

That is, the image is 25.5mm to the right of the vertex A.

(ii) For ophthalmic optics it is often preferable to think in vergences. Repeating the above and noting $L = \dfrac{n}{\ell}$, $L' = \dfrac{n'}{\ell'}$, and $F = \dfrac{n' - n}{r}$, we have

$$L' - L = F \quad \dotfill (1)$$

Putting all the distances in metres
$$L = \frac{n}{\ell} = \frac{1}{-0.30} = -3.33D, \text{ and } F = \frac{n' - n}{r} = \frac{4/3 - 1}{+0.006} = +55.55D.$$

Hence, substituting in equation (1),
$$L' = L + F = -3.33 + 55.55 = +52.22D,$$
whence $\ell' = \dfrac{n'}{L'} = \dfrac{4/3}{52.22} = +0.0255m \equiv +25.5mm.$

N.B. A slightly modified notation is used in some texts[*]. Since, $L = \frac{n}{\ell} = \frac{1}{\ell/n}$ it is possible to write $\frac{\ell}{n} = \bar{\ell}$ where $\bar{\ell}$ is read as 'ℓ reduced'. In keeping with this 'reduced' notation we can write

$$\bar{L} = \frac{1}{\bar{\ell}} \quad \& \quad \bar{L}' = \frac{1}{\bar{\ell}'} \; .$$

The bars over the letters serve to remind us that the quantities are reduced. Using this notation the refraction equation reads

$$\bar{L}' - \bar{L} = F.$$

Terms such as $1/\bar{\ell}$ are clearly curvatures whereas the equivalent n/ℓ has no such immediate physical significance. However, despite the advantage of the reduced notation it will not be used in this text.

(c) The object subtends 2^0 at the vertex. From figure 5.5

$$h = -\ell \tan 2^0 = -(-300 \times 0.0349) = +10.47\text{mm}.$$

The lateral or transverse magnification is given by

$$m = \frac{h'}{h} = \frac{L}{L'} \; .$$

Hence, $h' = h \cdot \frac{L}{L'} = 10.47 \times \frac{(-3.33)}{+52.22}$

$$= -0.667\text{mm}.$$

N.B. The minus sign indicates an inverted image.

[*] E.g. Technical Optics by L.C. Martin and W.T. Welford

---

5. Find the size of the image in the previous system if the object subtends 2^0 but is at a very large distance (at infinity).

The image size is given by

$$h' = f \tan \omega$$
$$= -18.0 \tan 2^0$$
$$= -18.0 \times 0.0349 = -0.628\text{mm}.$$

---

6. Show that when a cone of convergent rays is intercepted by a plane parallel plate of glass of thickness t and refractive index n_g, the position of the focus is moved outwards, away from the plate, by $t - \frac{t}{n_g}$.

In the absence of the glass plate the rays would focus at B_1. In the presence of the plate they focus at B_2'. Hence, the shift is $B_1 B_2'$, see figure 5.6 .

We are using $B_1, B_1', B_2,$ and B_2' to denote the object and image points for the first and second surfaces. Subscripts 1 and 2 refer to the first and second surfaces, and undashed and dashed letters refer to object and image points, respectively.

We shall require $B_1 B_2' = \ell_2' - (\ell_1 - t) = \ell_2' - \ell_1 + t$ ....(1)

We can use $L' = L + F$ at curved and plane surfaces, and indeed thin lenses, but in the case of plane surfaces $F = 0$ so $L' = L$.

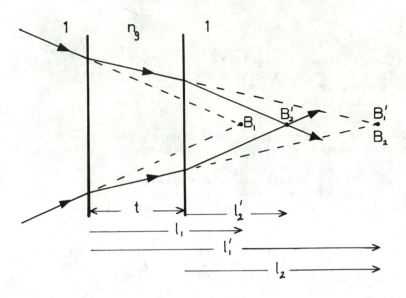

Fig. 5.6

Since $L' = \frac{n'}{\ell'}$ and $L = \frac{n}{\ell}$, $\frac{n'}{\ell'} = \frac{n}{\ell}$, or on rearranging $\ell = \frac{n}{n'}.\ell'$.

Using the latter at the first surface where $n_1 = 1$ and $n_1' = n_g$,

$$\ell_1 = \frac{n_1}{n_1'}.\ell_1' = \frac{\ell_1'}{n_g} \qquad \dots\dots\dots\dots\dots\dots(2)$$

Also, from the figure, $\ell_2 = \ell_1' - t \qquad \dots\dots\dots(3)$

At the second surface,

$$\ell_2' = \frac{n_2'}{n_2}.\ell_2 = \frac{n_2'}{n_2}(\ell_1' - t) = \frac{\ell_1'}{n_g} - \frac{t}{n_g} \qquad \dots\dots\dots(4)$$

Hence, using equations (2) and (4),

$$B_1B_2' = \ell_2' - \ell_1 + t$$

$$= \frac{\ell_1'}{n_g} - \frac{t}{n_g} - \frac{\ell_1'}{n_g} + t$$

$$= t - \frac{t}{n_g}$$

---

7. A concave spherical surface separating air and glass of
refractive index $n_g = 3/2$ has a power $-10.00D$. If the air
is replaced by water, find the power of the surface. The
water has refractive index $4/3$.

Let the power in air be $F_a = (n_g - 1)R$ and in water $F_w = (n_g - n_w)R$

where we have used the general relationship $F = \frac{n' - n}{r} = (n' - n)R$
where $R = 1/r$.

Now, $F_w/F_a = \dfrac{(n_g - n_w)R}{(n_g - 1)R}$

so, $F_w = \dfrac{(n_g - n_w)}{(n_g - 1)} \cdot F_a = \dfrac{(3/2 - 4/3)}{(3/2 - 1)}(-10) = -3\frac{1}{3}D.$

---

8. A contact lens has a back optic radius r. When in situ
on the eye this surface is bounded by the plastics material
and the tears. If the refractive indices of the plastics
material and the tears are n_p and n_t, respectively, show that
the power of this surface may be considered as two surfaces,
each of radius r, one a plastics/air boundary and the other
an air/tears boundary, the two being separated by an
infinitesimal layer of air.

In situ on the cornea the surface's power is
$$F = (n_t - n_p)R, \quad \text{where } R = \frac{1}{r}.$$

A mathematical trick allows us to express the power as follows:
$$F = (n_t - 1 + 1 - n_p)R$$
$$= (n_t - 1)R + (1 - n_p)R.$$

The first term represents the power of the air/tears surface
whilst the second term is the power of the plastics/air
surface. They are imagined separated by an infinitesimal
air layer so the vergence leaving the one is immediately
incident upon the other.

---

6. THIN LENSES

1. a) A virtual object is 20cm to the right of a thin lens
    of focal length +50cm. Find the position of the image.
    Assume the lens is in air.

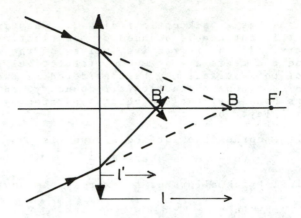

Fig. 6.1

Using $\frac{1}{\ell'} - \frac{1}{\ell} = \frac{1}{f'}$, we have $\frac{1}{\ell'} = \frac{1}{\ell} + \frac{1}{f'} = \frac{1}{+20} + \frac{1}{+50}$

$$= \frac{5 + 2}{100} = \frac{7}{100} .$$

Hence, $\ell' = +100/7 = +14.29$cm.

b) A real object is 50cm to the left of a thin diverging
lens of focal length $-33\frac{1}{3}$cm. If the object is 2cm high
find the position, size, and nature of the image.

We can always use the refraction equation $L' = L + F$ on
thin lenses. Thus
  $$L' = L + F = \frac{n}{\ell} + \frac{1}{f'} = \frac{1}{-0.5} + \frac{1}{-0.3} = -2 + (-3) = -5D.$$

Rearranging the expression $L' = \frac{n'}{\ell'}$, the image distance is

  $$\ell' = \frac{n'}{L'} = \frac{1}{-5} \text{ m} \equiv -20\text{cm}.$$

N.B. ℓ and f' must be in metres.

The image size may be found using the magnification
relationship, $m = \frac{h'}{h} = \frac{L}{L'}$,

  whence, $h' = h \cdot \frac{L}{L'} = 2 \cdot \frac{(-2)}{(-5)}$ $\text{cm} \cdot \frac{D}{D}$

      $= 0.8$cm.

The positive sign indicates an erect image. That the image
is virtual is determined by inspection of a diagram, see
figure 6.2, and the fact that L' is negative.

c) Rework the last two problems employing Newton's equations

$$xx' = -f'^2 \qquad \text{and} \qquad m = -\frac{f}{x} = -\frac{x'}{f'}.$$

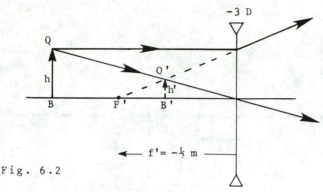

Fig. 6.2

(i) For the positive lens we refer to figure 6.3 . Remember
the extra-focal distances are measured from F to B, for x,
and from F' to B' for x', and the sign convention is applied.
So, FB = FO + OB = +(50 + 20) = +70 cm.

Hence, $x' = -\frac{f'^2}{x} = -\frac{50^2}{70} = -35.71$ cm;

i.e.   x'= F'B' = -35.71 cm, placing the image +14.29 cm from
the lens, as before.

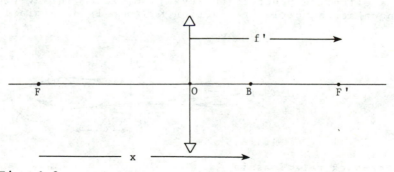

Fig. 6.3

(ii) Refer to figure 6.4 .

Again, $x' = -\dfrac{f'^2}{x} = -\dfrac{(33.33)^2}{-83.33} = +13.33$ cm, placing B' -20 cm from

the lens, by inspection.

For the image size we may use either $\dfrac{h'}{h} = -\dfrac{f}{x}$, or $\dfrac{h'}{h} = -\dfrac{x'}{f'}$.

Using the former, $h' = -\dfrac{f}{x} \cdot h = -\dfrac{(+33.33)}{(-83.33)} \times 2 = +0.8$ cm.

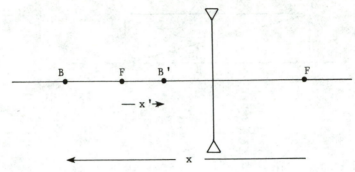

Fig. 6.4

2. An eye requires a vergence of -4.76 D incident upon the
    cornea in order to produce a clear image on the retina
    when the ciliary muscle is relaxed. What thin lens worn
    at 10 mm spectacle distance will allow this unaccommodated
    eye to see a distant object clearly?

The spectacle lens F must refract the incident light so that
the effective power at the eye is -4.76 D. This vergence is
known as the ocular refraction K, so F is determined by the

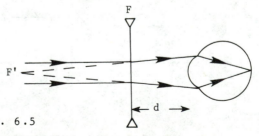

Fig. 6.5

step-back relationship

$$F = \dfrac{K}{1 + dK} = \dfrac{-4.76}{1 + 0.010 \times (-4.76)} = -5 \text{ D}.$$ Note that d must

be in metres so the term dK has units $m \times m^{-1}$; i.e. it is a
pure number with no units.

3. Suppose the spectacle lens in the last question is to be worn at 15mm. What power is required?

(i) We could proceed as before putting $d = 0.015$m in the equation $F = K/(1 + dK)$. However, suppose we write F_{new} for the power required at the new position, and F_{old} for the power in the original (old) position. Correspondingly, $d_{new} = 15$mm and $d_{old} = 10$mm. Now, $K = -4.76$D is constant for this eye and we can write, using the effectivity equation twice

$$K = \frac{F_{new}}{1 - d_{new} \cdot F_{new}} = \frac{F_{old}}{1 - d_{old} \cdot F_{old}} \quad .$$

Solving for F_{new} gives

$$F_{new} = \frac{F_{old}}{1 + (d_{new} - d_{old})F_{old}}$$

Hence, $F_{new} = \dfrac{-5}{1 + ((0.015 - 0.010)(-5))} = -5.128$D.

(ii) An approximate expression can be obtained using a binomial expansion and neglecting terms raised to the power of 2 or more.
Thus,
$$F_{new} = F_{old} \times \frac{1}{1 + (d_{new} - d_{old})F_{old}}$$

$$= F_{old}(1 - (d_{new} - d_{old})F_{old} + \dots + \text{higher orders})$$

$$\simeq F_{old} - (d_{new} - d_{old})F_{old}^2 \quad .$$

This expression says the 'new' spectacle lens differs from the 'old' spectacle lens by $-(d_{new} - d_{old})F_{old}^2$.
Let us apply this to the problem.

$$F_{old} = -5\text{D},$$
so, $d_{new} - d_{old} = 15 - 10 = 5\text{mm} \equiv 0.005\text{m}$.
Hence, $-(d_{new} - d_{old})F_{old}^2 = -0.005 \times (-5)^2 = -0.125 \quad \text{m.D}^2 (=\text{D})$.

This makes the new lens power , at 15mm, $-5 - 0.125 = -5.125$D which compares well with -5.128D by the exact equation.

Notes:
(i) The units $\text{m.D}^2 = \text{m.m}^{-2} = \text{m}^{-1} = \text{D}$ in the approximate expression. When deriving an equation we should always check the units are consistent.
(ii) The error in the approximate calculation, that is $-5.128 - (-5.125) = -0.003$D, is the sum of all the second and higher order terms in the infinite series expansion. This is very small and can clearly be ignored.
(iii) The binomial expansion of a term like $(1 + x)^{-1}$ is only possible when $|x| < 1$. In our particular case

$(d_{new} - d_{old})F_{old}$ is always likely to be less than unity since the spectacle distances are in metres and these are small quantities.   Their difference is even smaller!

---

4. A lamp and screen are 1m apart and a +4.5D lens is mounted between them. Where must the lens be placed in order to produce a sharp image on the screen, and what will be the magnification?

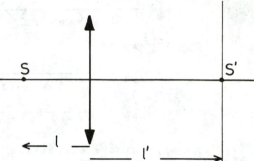

From the optics and the geometry of the situation we have,

$\frac{1}{\ell'} = \frac{1}{\ell} + \frac{1}{f'} = \frac{1}{\ell} + F$ ...(1)

and $\ell' - \ell = 1$ ......(2)

In the second equation we are saying that the distance $SS' = \ell' - \ell = 1m$.

Fig. 6.6

Note that $SS' \neq \ell' + \ell$ since ℓ, when given a value, will take a negative sign.

Rearranging equation (1),

$$\ell' = \frac{\ell}{1 + F\ell} \quad ....(3),$$

and rearranging equation (2),

$$\ell' = \ell + 1 \quad ....(4).$$

Equating the R.H.S. of equations (3) and (4),

$$\ell + 1 = \frac{\ell}{1 + F\ell} \ .$$

Rearranging this last equation,

$$(\ell + 1)(1 + F\ell) = \ell$$
$$\text{or,} \quad F\ell^2 + F\ell + 1 = 0 \ .$$

Since $F = +4.5$, we have

$$4.5\ell^2 + 4.5\ell + 1 = 0 \ .$$

This is a quadratic in ℓ. The solution of the general quadratic $ax^2 + bx + c = 0$ is

$$x = \frac{-b \pm \sqrt{(b^2 - 4ac)}}{2a}$$

and, using this for ℓ, where a=4.5, b=4.5, and c= 1,

$$\ell = \frac{-4.5 \pm \sqrt{((4.5)^2 - (4\times4.5\times1))}}{2 \times 4.5} = -\frac{1}{2} \pm \frac{1}{6} \text{ m}.$$

That is, the distance of the lamp from the lens has two possible values:

$$\ell = -\frac{1}{3}\text{m} \text{ , when } \ell' = \frac{2}{3}\text{m} ,$$

$$\ell = -\frac{2}{3}\text{m} \text{ , when } \ell' = \frac{1}{3}\text{m} .$$

The magnifications are ℓ'/ℓ, which gives -2 and $-\frac{1}{2}$ for the two positions.

Note: in general, magnification is L/L'

$$\text{and } \frac{L}{L'} = \frac{n/\ell}{n'/\ell'} = \frac{n}{\ell} \cdot \frac{\ell'}{n'}$$

For a thin lens in air $n = n' = 1$, so the magnification is $\frac{\ell'}{\ell}$.

---

5. Calculate the focal length of the lens required to produce an image of a given object on a screen with linear magnif -ication 3.5, the object being situated 20cm from the lens.

Magnification $m = \frac{\ell'}{\ell}$.

Hence, $-3.5 = \frac{\ell'}{-20}$,

or, $\ell' = (-20)\times(-3.5) = +70\text{cm}.$

The focal length is given by $\frac{1}{f'} = \frac{1}{\ell'} - \frac{1}{\ell}$

or, $f' = \frac{\ell\ell'}{\ell - \ell'} = \frac{(-20)\times70}{(-20)-70} = \frac{-1400}{-90} = +15.56\text{cm}.$

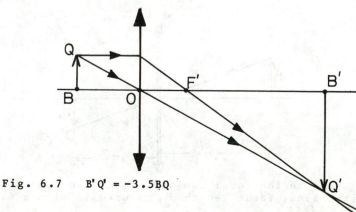

Fig. 6.7 $B'Q' = -3.5\,BQ$

6.   The difference in the positions of the image when the
     object is first at a great distance (at infinity) and
     then at 5m from the first focal point of a convex lens
     is 3mm. Find the focal length of the lens.

     When the object is at $-\infty$, $\ell' = f'$. In the second case,
     using Newton's relationship with $x = -5m$ and $x' = +0.003m$

$$xx' = -f'^2$$
     whence, $(-5) \times 0.003 = -f'^2$
        so, $f' = \sqrt{(0.015)} = 0.1225m \equiv 12.25cm.$

     Of course, in bringing the object in from $-\infty$ to $x=-5m$,
     the image moves from the second focal point a distance
     3mm to the right, making $x' = +0.003m$.

___

7.   A real, inverted image of an object is formed on a screen
     by a thin lens, and the image and object are equal in size.
     When a second thin lens is placed in contact with the first
     the screen must be moved 2cm nearer the lenses in order to
     obtain a clear image, and the size of this image is three-
     quarters that of the first image. Find the focal lengths
     of the two lenses.

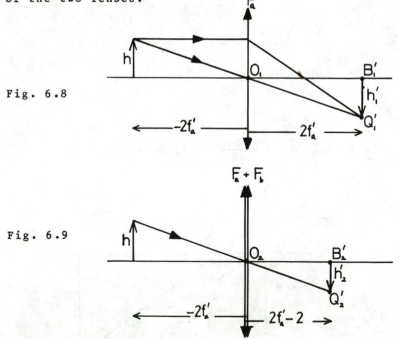

Fig. 6.8

Fig. 6.9

Let f'_a and f'_b be the focal lengths of the two lenses. The
The first lens, focal length f'_a, is used in figure 6.8 to

create the image equal in size to the object. The second
lens is combined with the first in figure 6.9 to produce
the image $\frac{3}{4}$ the size of the first image.
From the figures, triangles $O_1B_1O_1'$ and $O_2B_2O_2'$ are similar.
We are told that

$$h_2'/h_1' = \tfrac{3}{4} .$$

Also, in the first case, a thin positive lens produces an
inverted image equal in size to the object when $\ell' = -\ell = 2f'$.
Hence, the image distances in the two cases are $2f_a'$ and
$2f_a' - 2$, both distances in cm; see figures 6.8 and 6.9 .

Thus, using the two similar triangles

$$\frac{h_2'}{h_1'} = \frac{3}{4} = \frac{2f_a' - 2}{2f_a'} \quad , \text{ which, on rearranging, gives}$$

$$\frac{3}{4} \times 2f_a' = 2f_a' - 2 , \text{ and finally } f_a' = +4 \text{cm}.$$

For the combined lenses the object and image distances are
$-2f_a'$ and $2f_a' - 2$ cm, i.e. -8cm and $+6$cm. The power of the
combination is $F = F_a + F_b$, whence $\frac{1}{f'} = \frac{1}{f_a'} + \frac{1}{f_b'}$.

Using the thin lens equation $\frac{1}{\ell'} - \frac{1}{\ell} = \frac{1}{f'}$, we have

$$\frac{1}{6} - \frac{1}{-8} = \frac{1}{f_a'} + \frac{1}{f_b'} \quad , \text{ since } \frac{1}{f'} = \frac{1}{f_a'} + \frac{1}{f_b'} .$$

Now, $f_a' = +4$cm, so

$$\frac{1}{f_b'} = \frac{1}{6} - \frac{1}{-8} - \frac{1}{f_a'} = \frac{1}{6} + \frac{1}{8} - \frac{1}{4} = \frac{1}{24}$$

or, $f_b' = +24$cm.

8.   An image formed on a screen by a convex lens is 10cm long.
Without moving either the object or the screen, which are
300cm apart, a second image can be produced which is 40cm
long. Show how this is possible, and find the focal power
of the lens and the size of the object.

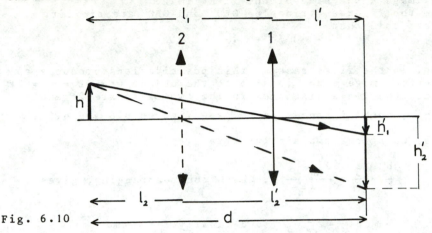

Fig. 6.10

Figure 6.10 shows the lens in position 1, with object and
image distances ℓ_1 and ℓ_1'. Since the light is reversible we
could have another object at $\ell_2 = -\ell_1'$ from the lens which
would produce an image at $\ell_2' = -\ell_1$ from the lens.

From the figure, $\ell_1' = d + \ell_1$ since ℓ_1 is negative.

Hence, $L_1' = \dfrac{n_1'}{\ell_1'} = \dfrac{1}{\ell_1'} = \dfrac{1}{d+\ell_1} = \dfrac{1}{d+\ell_1} \times \dfrac{L_1}{L_1} = \dfrac{L_1}{1+dL_1}$ ....(1)

where d is the distance between the object and the image.
If h is the object size and h_1' the image size when the lens
is in the first position, then

$$\frac{h_1'}{h} = \frac{L_1}{L_1'} = \frac{L_1}{L_1/(1+dL_1)} = 1+dL_1 \qquad ....(2)$$

Similarly, if h_2' is the image size with the lens in the
second position,

$$\frac{h_2'}{h} = \frac{L_2}{L_2'} = 1+dL_2 \qquad ....(3)$$

but $L_2 = \dfrac{n_2}{\ell_2} = \dfrac{1}{-\ell_1'} = -L_1' = \dfrac{-L_1}{1+dL_1}$ from equation (1) ....(4)

Substituting for L_2 from equation (4) into equation (3),

$$\frac{h_2'}{h} = 1+dL_2 = 1 + \frac{d(-L_1)}{1+dL_1} = \frac{1+dL_1-dL_1}{1+dL_1} = \frac{1}{1+dL_1}$$

which gives on taking reciprocals,

$$\frac{h}{h_2'} = 1+dL_1 \qquad ....(5)$$

Note that the R.H.S. of equations (2) and (5) are equal.

Equating the L.H.S.

$$\frac{h}{h_2^!} = \frac{h_1^!}{h} ,$$

or, $h^2 = h_1^! \times h_2^! = -10 \times (-40) = 400 \text{cm}^2$.

Therefore, $h = \sqrt{400} = \pm 20$cm. The object is upright so we take +20cm.

Since $\frac{h_1^!}{h} = \frac{-10}{20} = \frac{L_1^!}{L_1} = 1 + dL_1$ (equation (2))

$$L_1 = (\frac{h_1^!}{h} - 1)/d = (-10/20 - 1)/3 = -0.5D, \text{ since } d = 3m.$$

From equation (1), $L_1^! = \frac{L_1}{1 + dL_1} = \frac{-0.5}{1 + 3(-0.5)} = +1D$.

Hence, the power of the lens is $F = L' - L = 1 - (-0.5) = +1.5D$.

An alternative solution

Since $\ell_2^! = -\ell_1$ and $\ell_2 = -\ell_1^!$,

$$m_1 = \frac{h_1^!}{h} = \frac{\ell_1^!}{\ell_1} \quad \text{and} \quad m_2 = \frac{h_2^!}{h} = \frac{\ell_2^!}{\ell_2} = \frac{-\ell_1}{-\ell_1^!} = \frac{\ell_1}{\ell_1^!} ,$$

whence $m_1 m_2 = \frac{h_1^! h_2^!}{h^2} = \frac{\ell_1^!}{\ell_1} \times \frac{\ell_1}{\ell_1^!} = 1$.

Hence, $h^2 = h_1^! h_2^! = (-10)(-40) = 400 \text{cm}^2$,
       or $h = 20$cm, taking the positive root.

To find the power of the lens we now note that

$$\frac{\ell_2^!}{\ell_2} = \frac{h_2^!}{h_2} = \frac{-40}{20} = -2 \quad\quad \text{or } \ell_2^! = -2\ell_2 \quad \ldots\ldots\ldots\ldots(1)$$

$$\text{and} \quad \ell_2^! - \ell_2 = 300\text{cm} \quad\quad \ldots\ldots\ldots\ldots\ldots(2)$$

Substituting for $\ell_2^!$ from (1) into (2),

$$-2\ell_2 - \ell_2 = 300$$

so, $\quad\quad \ell_2 = \frac{-300}{3} = -100$cm.

Hence, $\ell_2^! = -2\ell_2 = -2(-100) = +200$cm

and $\quad F = \frac{1}{f'} = \frac{1}{\ell'} - \frac{1}{\ell} = \frac{1}{2.00} - \frac{1}{-1.00} = +1.50D$.

Note: ℓ and ℓ' were expressed in metres in the last step.

9.* A thin positive lens produces a real image of a real object. Show that the shortest possible distance between the object and image is $4f'$ and that this occurs when $\ell' = 2f'$ and $\ell = -2f'$, where f' is the focal length of the lens.

Let s be the distance between the object and the image. Thus, $s = \ell' - \ell$, since ℓ is negative.

From the thin lens equation

$$\frac{1}{\ell'} = \frac{1}{\ell} + \frac{1}{f'}, \qquad \text{or } \ell' = \frac{\ell f'}{\ell + f'} .$$

Hence, $s = \ell' - \ell = \dfrac{\ell f'}{\ell + f'} - \ell = \dfrac{-\ell^2}{\ell + f'}$

Differentiating s with respect to ℓ,

$$\frac{ds}{d\ell} = \frac{-2\ell(\ell + f') - (-\ell^2)}{(\ell + f')^2} = \frac{-\ell(\ell + 2f')}{(\ell + f')^2} .$$

$\dfrac{ds}{d\ell} = 0$ when ℓ or $(\ell + 2f')$ is zero.

The latter is physically meaningful in the thin lens equation: (try putting $\ell = 0$ in the equation!).
    Thus,
$$\ell + 2f' = 0, \quad \text{or } \ell = -2f' .$$
Putting this value for ℓ in the thin lens equation gives $\ell' = 2f'$. So,
$$s = \ell' - \ell = 2f' - (-2f') = 4f' .$$

A plot of s against ℓ is given in figure 6.11

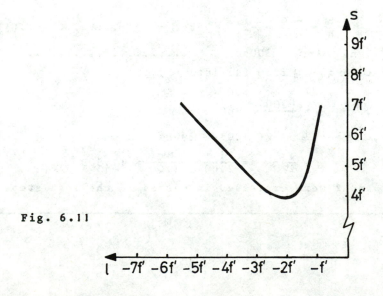

Fig. 6.11

10. The 'complete equation' for a thin lens in air is
$F = (n_g - 1)(R_1 - R_2)$, where n_g is the refractive index of
the glass and R_1 and R_2 are the curvatures of the first
and second surfaces. Object and image distances are
related to the lens by the usual equation

$$\frac{1}{\ell'} - \frac{1}{\ell} = \frac{1}{f'} .$$

What are the corresponding equations if the lens is
immersed in water, index n_w?

Since $F = F_1 + F_2$
and $\quad F_1 = (n_1' - n_1)R_1 \qquad$ and $\qquad F_2 = (n_2' - n_2)R_2$

$\qquad\qquad = (n_g - n_w)R_1 \qquad\qquad\qquad\quad = (n_w - n_g)R_2$

then $\quad F_w = (n_g - n_w)R_1 + (n_w - n_g)R_2$

$\qquad\qquad = (n_g - n_w)R_1 - (n_g - n_w)R_2$

$\qquad\qquad = (n_g - n_w)(R_1 - R_2)$, where F_w is the lens power

when immersed in water.
The refraction equation can be used for thin lenses.
Thus, $\qquad L' - L = F_w$

or, $\qquad \dfrac{n'}{\ell'} - \dfrac{n}{\ell} = F_w$

and $\qquad \dfrac{n_w}{\ell'} - \dfrac{n_w}{\ell} = (n_g - n_w)(R_1 - R_2)$

---

11. An object 2mm high is situated on the axis 25cm from a
+6D lens. Find the position, size, and nature of the image.
Show that if a cube of glass, edge 15cm and $n_g=1.5$, is
placed ANYWHERE along the axis between the object and the
lens the image distance doubles and the image size increases
by two and one-half times.

Without the glass block, we have

$$\frac{1}{\ell'} = \frac{1}{\ell} + \frac{1}{f'} = \frac{1}{\ell} + F = \frac{1}{-0.25} + 6 = +2D$$

Whence, $\quad \ell' = \dfrac{1}{+2}$ m $\equiv$ +50cm.

(Note: distances are in metres when mixing distances and
powers).
The image size is given by

$$h' = \frac{\ell'}{\ell}.h = \frac{50}{-25} \times 2mm = -4mm .$$

The image is inverted, the minus sign attached to its size

telling us this. It is a real image since the rays leaving the lens are convergent (L' = +2D).

If a glass block is placed between the object and the lens there will be three refracting elements in the system: two plane surfaces and the thin lens, figure 6.12 .

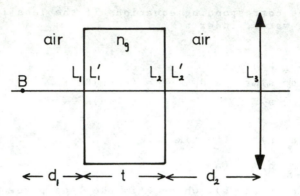

air $\quad n_g \quad$ air

$$B \quad L_1 \ L_1' \qquad L_2 \ L_2' \qquad L_3$$

$$\leftarrow d_1 \rightarrow \leftarrow t \rightarrow \leftarrow d_2 \rightarrow$$

Fig. 6.12

If we can show that the vergence arriving at the lens is independent of the block's position then we shall have achieved the required condition. We need to show that the air distances, d_1 and d_2, can take any values subject to the constraint that $d_1 + d_2 = 10$cm, in this case.

Tracing our way from the object to the lens,

$$L_1 = \frac{n_1}{\ell_1} = \frac{1}{-d_1} \qquad (d_1 \text{ and } d_2 \text{ positive})$$

$$L_1' = L_1 + F_1 = L_1 + 0 = \frac{1}{-d_1}, \text{ since } F_1 = 0 .$$

Using the step-along equation,

$$L_2 = \frac{L_1'}{1 - \frac{d}{n}L_1'} = \frac{\frac{1}{-d_1}}{1 - \frac{t}{n_g}(\frac{1}{-d_1})} = \frac{1}{-(d_1 + \frac{t}{n_g})} , \text{ after multiplying}$$

by $\frac{-d_1}{-d_1}$.

$$L_2' = L_2 + F_2 = \frac{1}{-(d_1 + \frac{t}{n_g})} , \text{ since } F_2 = 0$$

Stepping along to the lens,

$$L_3 = \frac{L_2'}{1 - \frac{d}{n}L_2'} = \frac{1/(-(d_1 + \frac{t}{n_g}))}{1 - d_2(1/-(d_1 + \frac{t}{n_g}))} = \frac{1}{-(d_1 + \frac{t}{n_g} + d_2)} .$$

We have here the proof which we require. We only need to interpret it. The numerator in the last expression is $n_3 = 1$, since the light reaching the lens is in air. The denominator is $\ell_3 = -(d_1 + t/n_g + d_2)$ where the term in the brackets is the sum of the air and air-equivalent distances from the object to the lens. Clearly, we may place the block of glass anywhere between the object and the lens providing $d_1 + d_2 = 10 cm$, which will always be true since the thickness of the glass is constant.

This analysis is the basis of the method used in question 6 in section 4.

Thus, $(d_1 + d_2) + \dfrac{t}{n_g} = 10 + \dfrac{15}{1.5} = 20 cm$. Expressing this in metres,

$$L_3 = \frac{1}{-0.20} = -5D$$

and $\quad L_3' = L_3 + F_3 = -5 + 6 = +1D.$

Hence, $\quad \ell_3' = \dfrac{n_3'}{L_3'} = \dfrac{1}{+1} = +1m \equiv +100 cm.$

The final image size, h_3', is given by

$$\frac{h_3'}{h_1} = m_1 \times m_2 \times m_3 = \frac{L_1}{L_1'} \times \frac{L_2}{L_2'} \times \frac{L_3}{L_3'} = \frac{L_3}{L_3'}$$

since $L_1 = L_1'$ and $L_2 = L_2'$.

Thus, $\quad h_3' = \dfrac{L_3}{L_3'} \cdot h_1 = \dfrac{-5}{+1} \times 2mm = -10mm.$

Inspection of the results shows that the image distance has doubled from +50cm to +100cm, and the image size has increased two and one-half times from -4mm to -10mm.

---

2. A thin positive lens of focal length f' is to cast a real image N-times larger that the object. Show that the image distance, ℓ', is equal to $(N+1)f'$.

The magnification $\quad m = \dfrac{\ell'}{\ell} = -N$,

whence $\quad \ell' = -N\ell \quad \dots\dots\dots\dots\dots\dots(1)$

From the thin lens equation,

$$\frac{1}{\ell} = \frac{1}{\ell'} - \frac{1}{f'} = \frac{f' - \ell'}{\ell' f'},$$

or, $\quad \ell = \dfrac{\ell' f'}{f' - \ell'} \quad \dots\dots\dots\dots\dots\dots(2)$

Substituting for ℓ from equation (2) into equation (1),

$$\ell' = -N \cdot \frac{\ell' f'}{f' - \ell'}$$

Dividing through by ℓ' and rearranging,

$$f' - \ell' = -Nf'$$

which gives $\ell' = (N+1)f'$.

---

13. An object 2cm high lies 50cm in front of a lens the front
surface of which is convex with radius 50cm. If a real
image 4cm high is produced by the lens, what is the radius
of the second surface? The refractive index of the lens
is $n_g = 1.5$.

The magnification $m = \dfrac{\ell'}{\ell} = \dfrac{h'}{h} = \dfrac{-4}{2} = -2$.

Hence, $\ell' = -2\ell = -2(-50) = +100\,cm$.

Now, $\dfrac{1}{f'} = \dfrac{1}{\ell'} - \dfrac{1}{\ell} = \dfrac{1}{100} - \dfrac{1}{-50} = \dfrac{3}{100}$

and $\dfrac{1}{f'} = \dfrac{3}{100} = (n_g - 1)(\dfrac{1}{r_1} - \dfrac{1}{r_2})$

from which

$$\dfrac{n_g - 1}{r_2} = \dfrac{n_g - 1}{r_1} - \dfrac{1}{f'} \quad .$$

Putting in the values,

$$\dfrac{1.5 - 1}{r_2} = \dfrac{1.5 - 1}{50} - \dfrac{3}{100} = \dfrac{-2}{100}$$

and $r_2 = \dfrac{1}{2}(-\dfrac{100}{2}) = -25\,cm$.

---

7.   REFLECTION AT SPHERICAL AND ASPHERICAL SURFACES

1. A concave spherical mirror of radius 200cm converges light
    from a distant object on to a concave mirror of radius 120cm.
    The latter is 40cm in front of the former, figure 7.1 . The
    light comes to a focus and is made parallel by a thin +10D
    lens. Where should this lens be placed?

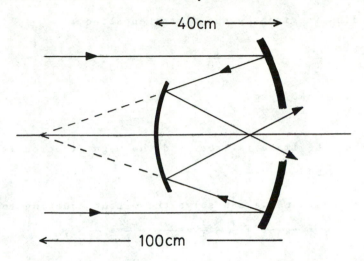

Fig. 7.1

The parallel incident light is focused at the focal point of
the 200cm radius mirror. Considering the smaller mirror, the
converging light provides a virtual object where $\ell = +60$cm.
The image distance is obtained from the mirror equation:

$$\frac{1}{\ell'} = \frac{1}{f'} - \frac{1}{\ell} = \frac{1}{-60} - \frac{1}{60} = -\frac{2}{60} = -\frac{1}{30} \ .$$

Hence, $\ell' = -30$cm. That is, the light focuses 30cm in front of
the mirror. But this is 10cm from the hole in the larger
mirror so a +10D lens placed in the 'plane of the hole' will
make the light parallel. The system is a little like a
Gregorian telescope.

2. A real inverted image twice the size of the object is produced 20cm from a mirror. What is the object's position when the magnification is four and the image is erect? Find the radius of curvature of the mirror.

Clearly, the mirror must be concave to produce a real image when the object is real. Since the image is inverted in the first case, we have

(i) $m = -2 = -\frac{\ell'}{\ell}$. Thus $\ell = \frac{1}{2}\ell' = \frac{1}{2} \times (-20) = -10\text{cm}$.

The radius is given by the mirror equation,

$$\frac{2}{r} = \frac{1}{\ell'} + \frac{1}{\ell} = \frac{1}{-20} + \frac{1}{-10} = -\frac{3}{20} .$$

That is, $r = -40/3$ cm.

(ii)

When the image is erect and the magnification is four we have
$$m = 4 = -\frac{\ell'}{\ell} .$$

Thus, $\ell' = -4\ell$ and we can solve the mirror equation for ℓ.

Hence, $\frac{1}{f'} = \frac{1}{-20/3} = \frac{1}{\ell'} + \frac{1}{\ell} = \frac{1}{-4\ell} + \frac{1}{\ell} = \frac{3}{4\ell}$,

or, $\frac{4}{3}\ell = -\frac{20}{3}$ and $\ell = -\frac{3}{4} \times \frac{20}{3} = -5\text{cm}$.

That is, the object is at -5cm from the mirror and the image is +20cm from the mirror. The two cases are drawn in figures 7.2 and 7.3 .

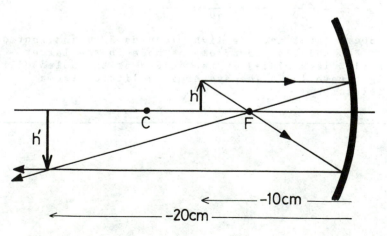

Fig. 7.2

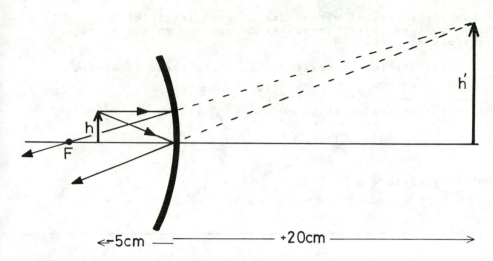

Fig. 7.3

---

3. An object and its image are respectively -20cm and +5cm
    from the centre of curvature of a concave mirror. Find
    the mirror's radius of curvature.

    The object distance is ℓ = r +(-20) = r-20 cm, since ℓ and
    r are negative, and the image distance is ℓ' = r +(+5) = r+5cm.

    Using the mirror equation, we have

    $$\frac{1}{\ell'} + \frac{1}{\ell} = \frac{2}{r}$$

    which on inserting the values for ℓ and ℓ' becomes

    $$\frac{1}{r+5} + \frac{1}{r-20} = \frac{2}{r} \ .$$

    Rearranging this equation,
    $$\frac{(r-20) + (r+5)}{(r+5)(r-20)} = \frac{2}{r}$$

    which becomes   r((r-20)+(r+5)) = 2(r+5)(r-20).
    Removing the brackets,
    $$r^2 - 20r + r^2 + 5r = 2r^2 + 10r - 40r - 200$$

    Cancelling the terms in r^2 and collecting other like terms
    then, $r = -\frac{200}{15}$ cm.

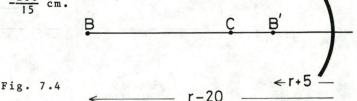

Fig. 7.4

4. A concave spherical mirror has a focal length of 10cm. Where must an object be placed to produce an erect image $1\frac{1}{2}$ times as large?

The magnification is positive since the image is erect. Hence, $m = -\frac{\ell'}{\ell} = \frac{3}{2}$. That is, $\ell' = -\frac{3}{2}\ell$.

Using the mirror equation, $\frac{1}{\ell'} + \frac{1}{\ell} = \frac{1}{f'}$, with $\ell' = -\frac{3}{2}\ell$,

$$\frac{1}{-\frac{3}{2}\ell} + \frac{1}{\ell} = \frac{1}{-10} ,$$

Multiplying through by 30ℓ ,

$$-20 + 30 = -3\ell , \quad \text{or} \quad \ell = -\frac{10}{3} \text{ cm}.$$

---

5. Show that the spherical mirror equation is applicable to a plane mirror.

The spherical mirror equation is $\frac{1}{\ell'} + \frac{1}{\ell} = \frac{1}{f'} = \frac{2}{r}$.

But $r = f' = \infty$ for a plane mirror and $\frac{1}{f'} = \frac{2}{r} = 0$.

Thus, $\frac{1}{\ell'} = -\frac{1}{\ell}$ or $\ell' = -\ell$, which says that the image is as far behind the plane mirror as the object is in front of it.

---

6. The Kitt Peak solar telescope consists of a plane mirror, 200cm across, which tracks the sun. It reflects light down a 150m shaft to a parabolic mirror which focuses the beam 100m back up the shaft. Find the size of the sun's image if the diameter of the sun is 1 382 400km and its distance from the Earth is 148 800 000km.

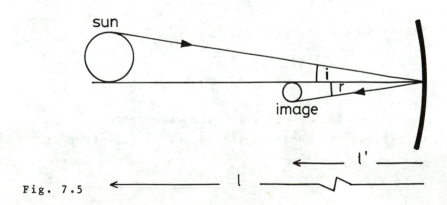

Fig. 7.5

Since i = r, figure 7.5, we have

$$i = r = \frac{\text{image size}}{|\ell'|} = \frac{\text{sun's diameter}}{|\ell|}$$

$$\therefore \quad \text{the image size} = \text{sun's diameter} \times \left|\frac{\ell'}{\ell}\right|$$

$$= 1\ 382\ 400 \times \frac{100 \times 10^{-3}}{148\ 800\ 000}$$

$$= 92.9\text{cm}.$$

---

7. A thin convex lens is used to provide a virtual object 5cm from the vertex of a convex mirror of focal length 10cm. If the real object for the lens is the same size as the virtual object for the mirror and the two objects are separated by 100cm, find the power of the lens, its position in relation to the mirror, and the relative size, position and nature of the final image.

A thin lens will produce a real inverted image the same size as the object, h, when $\ell' = -\ell = 2f'$. That is, the object and image are separated by 4f'. Thus, $4f' = 100$cm and $f' = +25$cm. The lens power is therefore $F = 1/f' = 1/+0.25 = +4D$.
The image formed by the lens is 50cm from the lens and 5cm behind the mirror. The lens is therefore 45cm in front of the mirror, figure 7.6 .

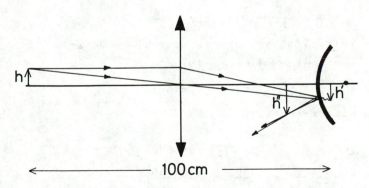

100 cm

Fig. 7.6

The object for the mirror, h' in the figure, is inverted and the object distance is +5cm. The final image distance is obtained from the mirror equation,

$$\frac{1}{\ell'} = \frac{1}{f'} - \frac{1}{\ell} = \frac{1}{10} - \frac{1}{5} = -\frac{1}{10} ;$$

i.e. $\ell' = -10$cm, or 10cm in front of the mirror.

The magnification is $m = \frac{h''}{h'} = -\frac{\ell'}{\ell} = -\frac{(-10)}{5} = 2$.

Hence, the final image, h", is inverted like the mirror's
object h'. Also, it is real and twice the size of the lens'
object, h.

---

8. A thin plano-convex lens of power +10D is silvered on
    its convex surface. If an object 2cm high is placed on
    the axis 100cm from the plane surface, find the position,
    size, and nature of the image produced by the lens-mirror.
    The refractive index of the lens is 1.5 .

    We shall need to know the radius of curvature of the
    mirrored surface. This can be obtained from the surface
    power equation $F = (n'-n)/r$.

    $$\text{Thus,} \quad r = (n'-n)/F = (1 - 1.5)/10 = -0.05m = -5cm \ .$$

    We shall consider a refraction at the plane surface, a
    reflection at the concave mirror, and a further refraction
    at a plane surface. When considering the second refraction
    the light will be travelling from right to left. There
    are several ways of tackling the problem and we shall
    consider two of them. They differ in the way we shall
    treat the reflection.

    a) Subscripts 1,2, and 3 apply to refraction at the plane
    surface, reflection, and refraction again.
    The vergence arriving at the plane surface is

    $$L_1 = \frac{n_1}{\ell_1} = \frac{1}{-1} = -1D, \text{ since } \ell_1 = -1m.$$

    Since the refracting surface is plane, $F_1 = 0$, and

    $$L_1' = L_1 + F_1 = -1 + 0 = -1D \ .$$

    Now, $\quad \ell_1' = \frac{n_1'}{L_1'} = \frac{1.5}{-1} = -1.5m \equiv -150cm.$

    Using the mirror equation,

    $$\frac{1}{\ell_2'} = \frac{2}{r} - \frac{1}{\ell_2} = \frac{2}{-5} - \frac{1}{-150} \quad , \text{ since } \ell_2 = \ell_1' \ ,$$

    and $\quad \ell_2' = \frac{150}{-60 + 1} = -\frac{150}{59} cm \equiv -\frac{1.50}{59} m.$

    The light reflected from the mirror is converging in glass
    to a point 1.50/59m to the left of the mirror. For the
    second refraction this converging light has an incident
    vergence of

    $$L_3 = \frac{n_3}{\ell_3} = \frac{1.5}{+1.50/59} = +59D \text{ (Note } \ell_3 = -\ell_2')$$

    so $\quad L_3' = L_3 + F_3 = 59 + 0 = +59D$

    and $\quad \ell_3' = \frac{n_3'}{L_3'} = \frac{1}{59} m \equiv +1.69cm.$

b) The reflection can be treated directly in vergences.
The vergence relationship $L' = L + F$ holds for a mirror
with the special conditions that

$$L' = -\frac{n'}{\ell'} \quad \text{and} \quad F = -\frac{n'}{f'} = -\frac{2n'}{r} .$$

The minus sign is necessary for the image vergence since
the light reverses direction. The minus sign in the
power equation comes about because a concave mirror
converges light.

$\quad L_2 = L_1' = -1D$. That is, the light leaving the
plane surface immediately strikes the mirror.

$$L_2' = L_2 + F_2 = -1 + (-\frac{n_2'}{f_2'}) = -1 + (\frac{-1.5}{-0.025}) = -1 + 60 = +59D$$

$\qquad$ (Note: the focal length is $-2.5cm \equiv -0.025m$).

At the second refraction,

$$L_3 = L_2' = +59D$$

and $\qquad L_3' = L_3 + F_3 = 59 + 0 = +59D.$

Finally, $\ell_3 = \frac{n_3'}{L_3'} = \frac{1}{59} m \equiv +1.69cm$.

In both cases the $+1.69cm$ indicates the image distance to
the left of the plane surface since the light is travelling
from right to left on meeting the plane surface for the
second time.

The magnification is given by

$$m = \frac{L_1}{L_1'} \cdot \frac{L_2}{L_2'} \cdot \frac{L_3}{L_3'} = \frac{(-1) \cdot (-1)}{(-1)} \cdot \frac{59}{59} = -\frac{1}{59}$$

which is just the magnification due to the mirror, L_2/L_2' .
The final image, h_3', is given by

$$h_3' = mh_1 = -\frac{1}{59} \times 2 = -\frac{2}{59} cm \text{ and it is inverted.}$$

9. A convex driving mirror is to be used to image the full
width of a road 10m wide at a distance 10¬ from the mirror.
The driver has to accommodate 2¬ to see the image clearly
when he is 40cm from the mirror. Find the position and
size of the image, and the radius of curvature and minimum
chord diameter of the mirror. Ignore the fact that the
driver is not on the principal axis of the mirror.

Since the driver accommodates 2D, the image is 50cm from
his eyes and therefore 10cm behind the mirror. The object
distance is -1000cm and the image size is given by the
magnification equation; thus

$$h' = mh = -\frac{\ell'}{\ell}.h = -\frac{10}{-1000}. \ 1000 = 10\text{cm}.$$

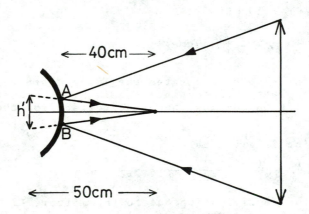

Fig. 7.7

From figure 7.7 the minimum chord diameter, AB, is given
by
$$AB = 40.\frac{h'}{50} = \frac{4}{5}.10 = 8\text{cm}.$$

That is, the mirror must be at least 8cm in diameter.
Since we know the object and image distances we can find
the radius of curvature.
Hence,

$$\frac{1}{\ell'} + \frac{1}{\ell} = \frac{2}{r}$$

which rearranges to give

$$r = \frac{2\ell\ell'}{\ell+\ell'} = \frac{2 \times (-1000) \times 10}{-1000 + 10} = +20.2\text{cm}.$$

8. CYLINDRICAL AND TOROIDAL LENSES AND SURFACES

1. A convex cylindrical surface has a radius of curvature
of 10cm and the refractive index of the glass is 1.7 .
What will be the reading on a lens measure, calibrated
for ophthalmic crown glass of index 1.523, when the three
legs are a) parallel to, b) perpendicular to, and c) at 30^0
to the cylinder axis? What would be the readings on a lens
measure calibrated for $n_g = 1.7$?

Let R_0, R_{30}, and R_{90} represent the curvatures at 0^0, 30^0, and
90^0 to the cylinder axis.
Then

$$\text{a)} \quad R_0 = \frac{1}{r_0} = \frac{1}{\infty} = 0 \; ,$$

$$\text{b)} \quad R_{90} = \frac{1}{r_{90}} = \frac{1}{+0.1} = +10D \; ,$$

$$\text{and} \quad \text{c)} \quad R_{30} \simeq R_{max} \cdot \sin^2 30^0 = 10 \times (\tfrac{1}{2})^2 = +2.5D \; .$$

Hence, if the glass were ophthalmic crown the powers
would be:

$$\text{a)} \quad F_0 = (n'-n)R_0 = (1.523 - 1) \times 0 = 0 \; ,$$

$$\text{b)} \quad F_{90} = (n'-n)R_{90} = (1.523 - 1) \times 10 = +5.23D,$$

$$\text{and} \quad \text{c)} \quad F_{30} = (n'-n)R_{30} = (1.523 - 1) \times 2.5 = +1.31D.$$

However, since the refractive index is 1.7 these figures
need correcting. The surface power equation applied for
glasses of refractive index 1.7 and 1.523 and with surface
curvature R gives

$$R = \frac{F_{1.7}}{1.7 - 1} = \frac{F_{1.523}}{1.523 - 1} \; .$$

Rearranging the R.H. equation,

$$F_{1.7} = \frac{1.7 - 1}{1.523 - 1} \cdot F_{1.523} = 1.338 \, F_{1.523} \; .$$

Hence, each of the power readings obtained previously
must be multiplied by a correcting factor of 1.338 .

Note: although the lens measure gives readings in meridians
other than the principal meridians the concept of power
does not apply. Rays in these other meridians do not focus
but undergo skew refraction.

2. A toroidal surface with principal radii of +8cm and +10cm
    in the horizontal and vertical meridians, respectively, is
    worked on one end of a glass rod 3cm in diameter. A luminous
    point source is embedded in the glass on the principal axis
    50cm from the surface. Find the positions and lengths of the
    focal lines and the position and diameter of the disc of
    least confusion. $n_g = 1.5$.

(i) The powers in the principal meridians are:

horizontally, $\quad F_H = \dfrac{n' - n}{r_H} = \dfrac{1.5 - 1}{8/100} = \dfrac{100 \times 0.5}{8} = +6.25D$ and

vertically, $\quad F_V = \dfrac{n' - n}{r_V} = \dfrac{1.5 - 1}{10/100} = \dfrac{100 \times 0.5}{10} = +5.00D$,

having here imagined the air to the left and glass to the
right of the surface.

(ii) The equations giving the positions of the focal lines
and the disc of least confusion are:

    dioptric position of 1st line $\quad = L_1'$

    linear position of 1st line $\quad = \ell_1' = n'/L_1'$

    dioptric position of 2nd line $\quad = L_2'$

    linear position of 2nd line $\quad = \ell_2' = n'/L_2'$

    dioptric position of disc of least confusion $= L_d' = (L_1' + L_2')/2$

    linear position of disc of least confusion $= \ell_d' = 2\ell_1'\ell_2'/(\ell_1' + \ell_2') = \dfrac{n'}{L_d'}$

Whether one uses the dioptric or the linear equations is a
matter of preference and the requirements of the question.
Certainly, when refracting the eye the practitioner always
thinks in dioptric terms. We shall use the dioptric equations
here.
The position of the first focal line is

$$L_1' = L_1 + F_H = \frac{n}{\ell_1} + F_H = \frac{1.5}{-0.50} + 6.25 = +3.25D$$

having turned the rod around so that the glass is to the left
of the surface. Note that the stronger principal meridian
is responsible for the first (nearer to the surface) focal
line.
Similarly, $\quad L_2' = L_2 + F_V = \dfrac{n}{\ell_2} + F_V = \dfrac{1.5}{-0.50} + 5.00 = +2.00D.$

and $\quad\quad\quad L_d' = \tfrac{1}{2}(L_1' + L_2') = \tfrac{1}{2}(3.25 + 2) = +2.625D$.

(iii) The dioptric equations for the focal line lengths and
the diameter of the disc of least confusion are:

length of first line = $D(\dfrac{L_1' - L_2'}{L_1'})$,

length of second line = $D(\dfrac{L_1' - L_2'}{L_2'})$,

diameter of disc of least confusion = $D(\dfrac{L_1' - L_2'}{L_1' + L_2'})$,

where D is the diameter of the refracting element.
Accordingly, the length of the first line is

$$3(\frac{3.25 - 2}{3.25}) = 1.15cm .$$

The length of the second line will be found to be 1.875cm
and the diameter of the disc of least confusion 0.71cm.

Notes: (i) The term inside the brackets has no dimensions.
       (ii) An alternative name for 'the disc of least
            confusion' is 'circle of least confusion'.

---

3. A thin lens, plano/+4.00DC x V, is to be made in toric
   equivalent with a minus 5D sphere curve. Find the radii
   of· curvature of the principal meridians of the toroidal
   surface.

Since the lens is thin the power is the sum of its surface
powers, i.e.
$$F = F_1 + F_2 \quad \dots\dots\dots\dots\dots\dots\dots\dots\dots\dots\dots\dots(1)$$

If we add δF to the front surface and subtract δF from
the back surface we do not change the power. In effect,
we are bending the lens, which is precisely what we want to
do.
Hence, rewriting equation (1), we have
$$F = (F_1 + \delta F) + (F_2 - \delta F) \quad \dots\dots\dots\dots\dots\dots(2)$$

The front surface of the plano-cylinder has powers $F_{1,V} = 0$
and $F_{1,H}$ = +4D in the vertical and horizontal meridians,
respectively. Considering these separately and noting that
F_2 is initially zero, the vertical power of the lens is
$$F_V = (F_{1,V} + 5) + (F_{2,V} - 5)$$

or, $0 = (0 + 5) + (0 - 5)$,

where the first term in brackets on the R.H.S. is the power
of the front surface of the toric in the vertical meridian
and the second term is the power of the back surface in the
same meridian. These are +5 and −5 respectively.

Considering the horizontal meridian,
$$F_H = (F_{1,H} + 5) + (F_{2,H} - 5)$$

or, $4 = (4 + 5) + (0 - 5)$,

making the toric front surface power +9 in the horizontal

meridian. The new F_2 is again -5 making the back surface
a -5D sphere as required. δF was 5, of course.

---

4. A point source and a screen are 1m apart. A thin sph/cyl
    lens -1.00/-2.00 x 90 is placed 25cm from the source. What
    lens placed 50cm from the source will produce a point image
    on the screen?

Figure 8.1 shows the set-up.

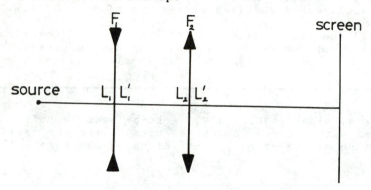

Fig. 8.1

Since the source is -0.25m from the first lens $L_1 = -4D$.
Similarly, $L_2' = +2D$ since the screen is +0.50m from the
second lens. In order to find the power of the second lens
we must find L_2 in both principal meridians, after which we
can find the power of the second lens in these meridians from
the equation $F = L' - L$.

In the horizontal meridian
$$L_1' = L_1 + F_1 = -4 + (-3) = -7D .$$

Using the step-along equation
$$L_2 = \frac{L_1'}{1 - \frac{d}{n}L_1'} = \frac{-7}{1 - \frac{0.25}{1}(-7)} = -2.55D ,$$

where d = +0.25m, the distance 'stepped along', and n = 1,
the medium 'stepped through'.
Hence,
$$F_2 = L_2' - L_2 = 2 - (-2.55) = +4.55D.$$

In the vertical meridian:
$$L_1' = L_1 + F_1 = -4 + (-1) = -5D ,$$

and $$L_2 = \frac{L_1'}{1 - \frac{d}{n}L_1'} = \frac{-5}{1 - \frac{0.25}{1}(-5)} = -2.22D ,$$

then $$F_2 = L_2' - L_2 = 2 - (-2.22) = +4.22D .$$

The lens required to make the beam stigmatic is therefore
+4.22DS/+0.33DC x 90. This is shown in terms of power
diagrams in figure 8.2 .

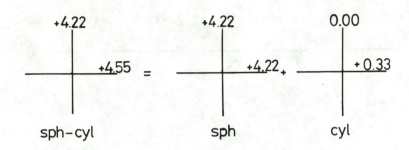

Fig. 8.2

---

5. A thin lens is made with a convex toroidal front surface
    and a plano back surface (a flat toric!). When a point
    source is on the axis at infinity it produces two line foci
    with a 2D interval of Sturm. The sum of the sags of the
    principal curvatures on the toroidal surface is 0.956mm
    when a lens measure with a 20mm separation between the outer
    legs is used. The refractive index of the glass is 1.523.
    Find the powers of the principal meridians.

Since all the power is due to the front surface we may
neglect the back surface. Labelling the principal meridian
powers F_1 and F_2 the difference between these is 2D. Let
F_1 be the stronger, then

$$F_1 - F_2 = 2 \quad \ldots\ldots\ldots\ldots\ldots\ldots\ldots\ldots\ldots(1)$$

Considering the sags, $s_1 + s_2 = 0.956$mm.

But $\quad s_1 \simeq \dfrac{y^2 F_1}{2000(n_g - 1)} \quad$ and $\quad s_2 \simeq \dfrac{y^2 F_2}{2000(n_g - 1)}$

so, $\quad s_1 + s_2 \simeq \dfrac{y^2}{2000(n_g - 1)}(F_1 + F_2)$

or, $\quad F_1 + F_2 \simeq 2000(n_g - 1)(s_1 + s_2)/y^2$

$$= 2000(1.523 - 1)(0.956)/10^2$$
$$= 10 \quad \ldots\ldots\ldots\ldots\ldots\ldots\ldots\ldots\ldots\ldots(2)$$

Adding equations (1) and (2),

$$2F_1 = 12$$

or, $F_1 = +6D$.

Substituting $F_1 = 6$ in equation (1) gives $F_2 = +4D$

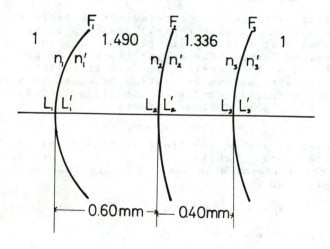

Fig. 9.1 This figure is for use with question 9.1 on page 55.

9.   SYSTEMS OF TWO OR MORE SURFACES OR THIN LENSES

1. Find the back vertex power(B.V.P. or F_v') and the equivalent power(F_E) of the following contact lens system in air.

   Outer lens radius............ 7.25mm.
   Inner lens radius............ 8.00mm.
   Final radius(liquid)......... 7.80mm.
   Contact lens refractive index 1.490, thickness 0.60mm.
   Liquid lens refractive index 1.336, thickness 0.40mm.

The B.V.P. is given by the vergence leaving the last surface, L_3', when the incident vergence at the first surface is zero. That is, $F_v' = L_3'$ when $L_1 = 0$. Accordingly, we must work our way through the system from left to right starting with $L_1 = 0$. To do this we shall need the surface powers:

$$F_1 = \frac{n_1' - n_1}{r_1} = \frac{1.490 - 1}{0.00725} = +67.59D$$

$$F_2 = \frac{n_2' - n_2}{r_2} = \frac{1.336 - 1.490}{0.0080} = -19.25D$$

$$F_3 = \frac{n_3' - n_3}{r_3} = \frac{1 - 1.336}{0.0078} = -43.08D$$

There are various ways of working through the system, all variations on a theme, and we shall stay in vergences. Using an inexpensive non-programmable calculator the arithmetic is very speedy. The scheme for calculation is then

   (i) find $L_1 = \frac{n_1}{\ell_1}$

   (ii) refract through the first surface, $L_1' = L_1 + F_1$

   (iii) step along to the second surface, $L_2 = \frac{L_1'}{1 - \frac{d}{n}L_1'}$,

      where d is the distance between the first and second surfaces (in metres) and n is the refractive index between them.
   (iv) refract through the second surface, and so on.

So, $L_1 = \frac{n_1}{\ell_1} = \frac{1}{-\infty} = 0$,

   $L_1' = L_1 + F_1 = 0 + 67.59 = +67.59D$

   $L_2 = \frac{L_1'}{1 - \frac{d}{n}L_1'} = \frac{67.59}{1 - \frac{0.0006}{1.490} \times 67.59} = +69.48D$.

Note that d must be in metres and $\frac{d}{n}L_1'$ must be calculated first and subtracted from 1 in the denominator.

   $L_2' = L_2 + F_2 = 69.48 + (-19.25) = +50.23D$.

$$L_3 = \frac{L_2'}{1-\frac{d}{n}L_2'} = \frac{50.23}{1-\frac{0.0004}{1.336}\times 50.23} = +51.00D$$

$$F_v' = L_3' = L_3 + F_3 = 51 + (-43.08) = +7.92D \simeq +8D \ .$$

The equivalent power of a system of r surfaces with air as the first and last medium is the power of that thin lens in air which will produce the same image size as the system. The image size for a distant object subtending an angle θ at the system is given by

$$f_1\tan\theta \cdot \frac{L_2}{L_2'} \cdot \frac{L_3}{L_3'} \cdots \frac{L_r}{L_r'} \quad \cdots\cdots\cdots\cdots(1)$$

The term $f_1\tan\theta$ is the image size produced by the first surface (or thin lens) and this is magnified at successive surfaces(or thin lenses). The terms L/L' are the magnifications at the refracting elements denoted by the subscripts they carry.
This final image size is equated to the image produced by an equivalent thin lens, $f_E\tan\theta$ ....(2)

Equating (1) and (2), and cancelling the $\tan\theta$ term,

$$f_E = f_1 \cdot \frac{L_2}{L_2'} \cdot \frac{L_3}{L_3'} \cdots\cdots \frac{L_r}{L_r'} \ .$$

Multiplying both sides of the equation by -1, applying the reciprocal function to both sides, and noting that $F_1 = -\frac{n_1}{f_1} = -\frac{1}{f_1}$, and $F_E = -\frac{1}{f_E}$, since air is to the left of the first refracting element, we have

$$F_E = F_1 \cdot \frac{L_2'}{L_2} \cdot \frac{L_3'}{L_3} \cdots \frac{L_r'}{L_r} \ .$$

In our particular problem

$$F_E = F_1 \cdot \frac{L_2'}{L_2} \cdot \frac{L_3'}{L_3} = 67.59 \times \frac{50.23}{69.48} \times \frac{7.92}{51.00} = +7.59D \ .$$

Notes:(i) In general, that is when the first and last media are not the same, the equivalent power represents an equivalent single surface. It represents an equivalent thin lens only when the first and last media are equal.

(ii) Generally, $F_E = -\frac{n_1}{f_E} = \frac{n_r'}{f_E'}$.

When $n_1 = n_r'$, we see that $f_E = -f_E'$.

2. A 3cm thick plano-convex lens is made of glass of refractive
    index n_g = 1.5 with the front surface +6.00D in air. If it
    is placed in water, n_w = 1.33, find the front and back vertex
    powers, the equivalent(principal) power, and the positions
    of the focal, principal, and nodal points.

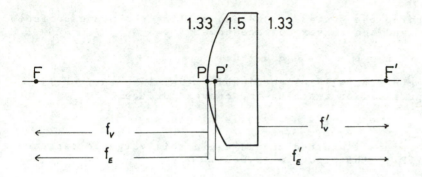

Fig. 9.2

(i) The power of the first surface in air is given by

$$F_{1,a} = \frac{n_1' - n_1}{r_1} = \frac{1.5 - 1}{r_1} = 6$$

so $$r_1 = \frac{1.5 - 1}{6} = +\frac{0.5}{6} \text{ m }.$$

The power of the surface when the lens is immersed in water
is

$$F_{1,w} = \frac{n_1' - n_1}{r_1} = \frac{1.5 - 1.33}{0.5/6} = \frac{6}{0.5} \times 0.17 = +2.04D .$$

F_2 = 0 both in air and in water.

(ii) Tracing parallel incident light through the system,
striking the convex surface first, will find the B.V.P.
since $L_2' = F_v'$. Thus, since L_1 = 0,

$$L_1' = L_1 + F_1 = 0 + 2.04 = +2.04D$$

$$L_2 = \frac{L_1'}{1 - \frac{d}{n}L_1'} = \frac{2.04}{1 - \frac{0.03}{1.5} \times 2.04} = +2.127D ,$$

$$L_2' = L_2 + F_2 = 2.127 + 0 = +2.127D = F_v' .$$

Now, denoting the back vertex by A_2,

$$A_2F' = f'_v = \frac{n'_2}{F'_v} = \frac{1.33}{2.127} = +0.6253m \equiv +62.53cm.$$

That is, the second principal focal point is 62.53cm to the right of the second surface.

(iii) Turning the system around and calling the plane surface power F_1 and the convex surface power F_2,

$$L'_1 = L_1 + F_1 = 0 + 0 = 0,$$

$L_2 = 0$, since parallel light leaving the plane surface is still parallel on reaching the convex surface.

$$L'_2 = L_2 + F_2 = 0 + 2.04 = +2.04D$$

When the lens is turned around again this vergence represents the F.V.P.; that is, $F_v = +2.04D$.

Denoting the vertex of the first surface, the convex surface, by A_1, the first principal focal point is

$$A_1F = f_v = -\frac{n_1}{F_v} = -\frac{1.33}{2.04} = -0.652m \equiv -65.2cm$$

from the vertex of the convex surface (to the left).

(iv) The equivalent focal lengths are

$$f_E = f_1 \cdot \frac{L_2}{L'_2} = -\frac{n_1}{F_1} \cdot \frac{L_2}{L'_2} = -\frac{1.33}{2.04} \times \frac{2.127}{2.127}m = -65.2cm$$

and $$f'_E = -\frac{n'_2}{n_1} \cdot f_E = -\frac{1.33}{1.33} \times (-65.2)m = +65.2cm .$$

The equivalent power is
$$F_E = -\frac{n_1}{f_E} = \frac{n'_2}{f'_E} = \frac{1.33}{0.652} = +2.04D$$

(v) The distance of the second principal point, P', from the vertex of the second surface, A_2, is

$$A_2P' = f'_v - f'_E = 62.53 - 65.2 = -2.67cm$$

or, alternatively,

$$A_2P' = e' = -n'_2 \frac{t}{n_g} \cdot \frac{F_1}{F_E} = -1.33 \times \frac{3}{1.5} \times \frac{2.04}{2.04} = -2.66cm.$$

The 0.01cm discrepancy in the two calculations of e' is due to rounding errors.

The distance of the first principal point, P, from the vertex of the first surface, A_1, is

$$A_1P = f_v - f_E = -65.2 - (-65.2) = 0$$

or, again, $A_1P = e = n_1 \dfrac{t}{n_g} \cdot \dfrac{F_2}{F_E} = 1.33 \times \dfrac{3}{1.5} \times \dfrac{0}{2.04} = 0$

That is, P lies at the vertex of the convex surface, which is always the case for a plano-convex lens.

(vi) Since $n_1 = n_2'$, the nodal points N and N' lie at P and P'.

Note: if $n_1 \neq n_r'$ in a system with r surfaces, the nodal points are found from the relationships

$$FN = P'F' = f_E'$$
and
$$N'F' = FP = -f_E$$

---

3. An object, 2cm high, is placed 2.66m from the lens in question 2 (the lens being surrounded by water). Find the position and size of the image using the three methods: (i) ray tracing with refraction and step-along, (ii) the equation

$$\frac{n'}{\ell'} - \frac{n}{\ell} = F_E$$

where ℓ and ℓ' are measured from P and P', respectively, (iii) Newton's relationship $xx' = f_E f_E'$.

The magnification relationships to be used in each method are

(i) $m = \dfrac{L_1}{L_1'} \cdot \dfrac{L_2}{L_2'}$,

(ii) $m = \dfrac{L}{L'} = \dfrac{n_1}{\ell} \cdot \dfrac{\ell'}{n_2'}$,

(iii) $m = -\dfrac{f_E}{x} = -\dfrac{x'}{f_E'}$.

Method (i): refraction and step-along

We first find L_1 :

$$L_1 = \frac{n_1}{\ell_1} = \frac{1.33}{-2.66} = -0.50D \ ,$$

$$L_1' = L_1 + F_1 = -0.50 + 2.04 = +1.54D \ ,$$

$$L_2 = \frac{L_1'}{1-\frac{d}{n}L_1'} = \frac{1.54}{1-\frac{0.03}{1.5} \times 1.54} = +1.589D$$

$$L_2' = L_2 + F_2 = 1.589 + 0 = +1.589D$$

The final image position is given by finding the image distance measured from the vertex of the second surface:

$$\ell_2' = \frac{n_2'}{L_2'} = \frac{1.33}{1.589} = +0.837m \equiv +83.7cm \ .$$

That is, the image is 83.7cm to the right of the second surface. Since L_2' is positive the rays converge to form a real image.

The magnification is given by

$$\frac{h_2'}{h_1} = m = \frac{L_1}{L_1'} \cdot \frac{L_2}{L_2'} \ ,$$

whence, the final image size is

$$h_2' = h_1 \cdot \frac{L_1}{L_1'} \cdot \frac{L_2}{L_2'} = 2 \times \frac{(-0.50)}{1.54} \times \frac{1.589}{1.589} = -0.649cm.$$

The image is inverted and diminished.

Method (ii)

The object is $-2.66m$ from the first principal point since the latter is at the vertex of the first surface. Thus $\ell = -2.66m$. The image will be a distance ℓ' from the second principal point.

Then,

$$\frac{n_2'}{\ell'} - \frac{n_1}{\ell} = F_E \ ,$$

and

$$\frac{1.33}{\ell'} = \frac{n_1}{\ell} + F_E = \frac{1.33}{-2.66} + 2.04 = +1.54D \ .$$

Hence, $\ell' = \frac{1.33}{1.54} = +0.8636m \equiv +86.36cm.$

Since P' is $-2.66cm$ from A_2 the image is $86.36 - 2.66 = +83.7cm$ from A_2, as by method (i).

The image size is

$$h' = h \cdot \frac{n_1}{\ell} \cdot \frac{\ell'}{n_2'} = 2 \times \frac{1.33}{(-2.66)} \times \frac{0.8636}{1.33} = -0.649 \, cm.$$

As it should, this agrees with the first method's value.

Method (iii)

To find the image position we rearrange the equation $xx' = f_E \, f_E'$ to give

$$x' = \frac{f_E \, f_E'}{x} = \frac{(-65.2)(65.2)}{-(266 - 65.2)} = +21.17 \, cm.$$

Note: $x = \ell_1 - f_v = -266 \, cm - (-65.2 \, cm) = -200.8 \, cm.$

The image is ℓ_2' from the second surface and

$$\ell_2' = f_v' + x' = 62.53 + 21.17 = +83.7 \, cm, \text{ once again.}$$

The image size is given by

$$h' = -h \cdot \left(\frac{f_E}{x}\right) = -2 \times \left(\frac{-65.2}{-200.8}\right) = -0.649 \, cm.$$

---

4. The angular size of the sun is 0.5°. Find the diameter of its image formed on a screen by the lens whose front and back surfaces are $F_1 = +3D$ and $F_2 = +2D$. The centre thickness is 1cm and the refractive index is 1.5 .

Method (i)

The final image size, h_2', is given by

$$h_2' = h_1' \times m_2 = f_1 \tan\theta \times \frac{L_2}{L_2'},$$

where θ is the angular subtent of the distant object. f_1 is given by $-\frac{n_1}{F_1}$, but we need to calculate L_2 and L_2'. Starting with $L_1 = 0$, since the sun is effectively at infinity,

$$L_1' = L_1 + F_1 = 0 + 3 = +3D$$

$$L_2 = \frac{L_1'}{1 - \frac{d}{n} L_1'} = \frac{3}{1 - \frac{0.01}{1.5} \times 3} = +3.06D$$

$$L_2' = L_2 + F_2 = 3.06 + 2 = +5.06D$$

Hence, $h_2' = f_1 \tan\theta \times \frac{L_2'}{L_2} = -\frac{n_1}{F_1} \tan\theta \times \frac{L_2'}{L_2}$

$$= -\tfrac{1}{3} \times 0.008727 \times \frac{3.06}{5.06} = -1.759 \times 10^{-3} \text{m} \equiv -1.759 \text{mm}.$$

Method (ii)

If we find the equivalent focal length, f_E', the image size for a distant object is then given by
$$h' = -f_E' \tan\theta \ .$$

Accordingly,

$$f_E' = \frac{1}{F_E} = \frac{1}{F_1 + F_2 - \frac{t}{n}F_1F_2} = \frac{1}{3 + 2 - \frac{0.01}{1.5} \times 3 \times 2} = +0.2016\text{m}$$

$$\equiv 201.6\text{mm}.$$

Hence, $h' = -f_E' \tan\theta = -201.6 \times 0.008727 = -1.759\text{mm}$,
where $\tan\theta = \tan 0.5^0 = 0.008727$.

---

5. Bending a lens causes the image size to increase. To illustrate this bend the lens in question 4 so that the back surface is -3D and the back vertex focal length is the same. The latter is necessary in order to leave the screen in the same place. You will, of course, need to calculate the new F_1 and this can be done using the standard equation
$$F_1 = \frac{F_v' - F_2}{1 + \frac{t}{n}(F_v' - F_2)}$$

Image the sun on the screen again.

In question 4 the B.V.P. was $F_v' = L_2' = +5.06\text{D}$. Hence, the new F_1 is

$$F_1 = \frac{F_v' - F_2}{1 + \frac{t}{n}(F_v' - F_2)} = \frac{5.06 - (-3)}{1 + \frac{0.01}{1.5}(5.06 - (-3))} = \frac{8.06}{1 + (\frac{0.01}{1.5} \times 8.06)}$$

$$= +7.649\text{D} \ .$$

The equivalent focal length is

$$f'_E = \frac{1}{F_E} = \cfrac{1}{F_1 + F_2 - \frac{t}{n}F_1F_2} = \cfrac{1}{7.649 + (-3) - \frac{0.01}{1.5} \times 7.649 \times (-3)}$$

$$= +0.2082m$$
$$\equiv +208.2mm$$

The image size is

$$-f'_E \tan\theta = -208.2 \times \tan 0.5^0 = -208.2 \times 0.008727 = -1.817mm.$$

This is a 3% increase in image size. Such use of the form of the lens in clinical practice is dealt with in Introduction to Visual Optics, by A.H. Tunnacliffe.

---

6. A gypsy's glass ball, diameter 10cm, is used to focus the sun's rays. They focus 2.5cm from the surface. Find the refractive index of the glass, n_g.

Using the Gaussian equation $\frac{n'}{\ell'} - \frac{n}{\ell} = \frac{n'-n}{r}$ at each surface we can arrive at an equation in one unknown, n_g.

At the first surface

$$\frac{n'_1}{\ell'_1} - \frac{n_1}{\ell_1} = \frac{n'_1 - n_1}{r_1},$$

where $\ell_1 = -\infty$, so $\frac{n_1}{\ell_1} = \frac{1}{-\infty} = 0$, and $n'_1 = n_g$.

Hence, $\dfrac{n_g}{\ell'_1} = \dfrac{n_g - 1}{5}$, since $r_1 = +5cm$,

and $\ell'_1 = \dfrac{5n_g}{n_g - 1}$.

At the second surface

Now, $\ell_2 = \ell'_1 - 10 = \dfrac{5n_g}{n_g-1} - 10 = \dfrac{10 - 5n_g}{n_g - 1}$,

$r_2 = -5cm$, $n_2 = n_g$, $n'_2 = 1$.

So, $\dfrac{n'_2 - n_2}{r_2} = \dfrac{1 - n_g}{-5} = \dfrac{n_g - 1}{5}$.

Using $\ell_2' = +2.5\text{cm}$, $\ell_2 = \dfrac{10 - 5n_g}{n_g - 1}$, and $\dfrac{n_2' - n_2}{r_2} = \dfrac{n_g - 1}{5}$ in the Gaussian

equation, we have

$$\frac{1}{2.5} - \frac{n_g}{(10 - 5n_g)/(n_g - 1)} = \frac{n_g - 1}{5} \ .$$

Multiplying through by $(10 - 5n_g)$ and rearranging the second term on the L.H.S.,

$$(4 - 2n_g) - n_g(n_g - 1) = (2 - n_g)(n_g - 1)$$

or, $4 - 2n_g - n_g^2 + n_g = 2n_g - n_g^2 - 2 + n_g$,

which reduces to $4n_g = 6$, or $n_g = 1.5$.

An alternative method

Note that parallel light is incident on the first surface so $\ell_2' = f_v' = +0.025\text{m}$.

Hence, $F_v' = \dfrac{1}{f_v'} = \dfrac{1}{+0.025} = +40\text{D}$.

Also, $F_1 = \dfrac{n_1' - n_1}{r_1} = \dfrac{n_g - 1}{0.05} = \dfrac{100(n_g - 1)}{5} = 20(n_g - 1)$

and $F_2 = \dfrac{n_2' - n_2}{r_2} = \dfrac{1 - n_g}{-0.05} = \dfrac{n_g - 1}{0.05} = F_1 = 20(n_g - 1)$,

and $t = 0.01\text{m}$.

Thus, $F_v' = \dfrac{F_1 + F_2 - \dfrac{t}{n_g}F_1F_2}{1 - \dfrac{t}{n_g}F_1}$, or $F_v'\left(1 - \dfrac{t}{n_g}F_1\right) = F_1 + F_2 - \dfrac{t}{n_g}F_1F_2$.

Multiplying both sides by n_g, and removing the brackets,

$$n_g F_v' - t F_v' F_1 = n_g F_1 + n_g F_2 - t F_1 F_2 \ .$$

Putting $F_v' = 40$, $F_1 = F_2 = 20(n_g - 1)$,

$40n_g - 0.10 \times 40 \times 20(n_g - 1) = 20n_g(n_g - 1) + 20n_g(n_g - 1) - 0.10 \times 20^2(n_g - 1)^2$.

Dividing both sides by 20 and removing the brackets,

$$2n_g - 4n_g + 4 = 2n_g^2 - 2n_g - 2n_g^2 + 4n_g - 2 \ ,$$

which gives $n_g = 1.5$, again.

7. The axial thickness of a lens is 5cm, and the vertex focal
   lengths are $f_v = -2$cm and $f_v' = +3$cm. When an object is 4cm
   from the front vertex the real image formed is twice the
   size of the object. Find the positions of the principal
   points if the lens is in air.

   If we can find the equivalent focal lengths, f_E and f_E', where
   $f_E = -f_E'$, we can find the positions of P and P'. Newton's
   equation, $m = -f_E/x$, will allow us to do this.

   Since the image is real, $m = -2$, and x is given by implication.
   $$x = \ell_1 - f_v = -4 - (-2) = -2\text{cm}.$$

   Thus, $f_E = -mx = -(-2)(-2) = -4$cm; i.e. $PF = f_E = -4$cm,
   which places P 2cm to the right of the first vertex, A_1,
   since $e = A_1P = f_v - f_E = -2 - (-4) = +2$cm.

   Now, $f_E' = -f_E = -(-4) = +4$cm,

   and $e' = A_2P' = f_v' - f_E' = 3 - 4 = -1$cm. That is, P' is 1cm to the
   left of the second vertex, A_2.

   ---

8. In the last question, find the surface powers if the
   refractive index of the lens is $n_g = 1.5$.

   We know the equivalent power and the vertex powers. They
   are

   $$F_E = \frac{n_2'}{f_E'} = \frac{1}{f_E'} = \frac{1}{+0.04} = +25\text{D},$$

   $$F_v' = \frac{n_2'}{f_v'} = \frac{1}{f_v'} = \frac{1}{+0.03} = +33.33\text{D},$$

   and $\quad F_v = -\frac{n_1}{f_v} = -\frac{1}{f_v} = -\frac{1}{(-0.02)} = +50\text{D}.$

   We can find F_1 either from the B.V.P. equation $F_v' = \dfrac{F_E}{1 - \frac{t}{n}F_1}$,

   or from $\quad A_2P' = -\dfrac{t}{n_g} \cdot \dfrac{F_1}{F_E}$. The latter looks the simpler: thus

   $$F_1 = -e'n_gF_E/t = -(-1) \times 1.5 \times 25/5 \qquad \text{cm.D/cm}$$

   $$= +7.5\text{D}.$$

Similarly, using $e = \dfrac{t}{n_g} \cdot \dfrac{F_2}{F_E}$, where $A_1P = +2cm$,

$$F_2 = e n_g F_E / t = 2 \times 1.5 \times 25 / 5 = +15D$$

Note: e and t must have the same units.

---

9. A telephoto lens in a camera consists of a +5D lens separated
    by 15cm from a -10D lens, the latter being nearer the film.
    Show that this is equivalent to a thin lens of focal length
    +40cm. What is the advantage of such a system compared to a
    camera with a single lens of focal length +10cm, say?

    The focal length of the equivalent thin lens is $f'_E = \dfrac{1}{F_E}$,

    where $F_E = F_1 + F_2 - \dfrac{t}{n} F_1 F_2 = F_1 + F_2 - d F_1 F_2$, since the medium between
    the lenses is air$(n = 1)$.   d is the distance (in metres)
    between the lenses.

    Now $F_E = 5 + (-10) - 0.15 \times 5 \times (-10) = +2.5D$,

    so $f'_E = \dfrac{1}{F_E} = \dfrac{1}{+2.5} m \equiv +40cm$.

    Since the image size is given by $-f' \tan\theta$, for a thin lens
    of focal length f' and a distant object subtending an angle θ,
    the image size is proportional to f'. Thus, if the focal
    length of a camera lens is 10cm, compared with the above
    telephoto system the latter produces an image 40cm/10cm = 4
    times as large .

    That is, if h' is the image size with a single thin lens and
    h'_t is the image size with the telephoto lens, then

    $$\dfrac{h'_t}{h'} = \dfrac{-f'_E \tan\theta}{-f' \tan\theta} = \dfrac{f'_E}{f'} = \dfrac{+40}{+10} = 4 \ .$$

    This principle is used in conjunction with the eye in some
    subnormal visual aids. The system is essentially a Galilean
    telescope.

---

10. A biconcave lens has front and back radii of 20cm and 10cm,
    respectively, and axial thickness 5cm. Describe the image
    of a 1cm tall object placed 12.5cm from the first vertex.
    $n_g = 1.5$.

    Using the refraction and step-along method, we first calculate
    the surface powers:

$$F_1 = \frac{n_1' - n_1}{r_1} = \frac{1.5 - 1}{-0.20} = -2.5D$$

and $\quad F_2 = \frac{n_2' - n_2}{r_2} = \frac{1 - 1.5}{0.10} = -5D$.

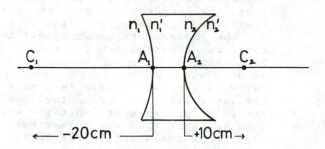

Fig. 9.3

Hence, $\quad L_1 = \frac{n_1}{\ell_1} = \frac{1}{-0.125} = -8D$

$$L_1' = L_1 + F_1 = -8 + (-2.5) = -10.5D$$

$$L_2 = \frac{L_1'}{1 - \frac{d}{n}L_1'} = \frac{-10.5}{1 - (\frac{0.05}{1.5} \times (-10.5))} = -7.78D$$

and $\quad L_2' = L_2 + F_2 = -7.78 + (-5) = -12.78D$,

whence, $\quad \ell_2 = \frac{n_2'}{L_2'} = \frac{1}{-12.78} = -0.0782m \equiv -7.82cm$.

That is, the image is 7.82cm to the left of A_2, or 2.82cm in front of A_1.
The final image size, h_2', is given by the magnification equation

$$\frac{h_2'}{h_1} = m_1 \times m_2 = \frac{L_1}{L_1'} \times \frac{L_2}{L_2'} \quad ,$$

or, $\quad h_2' = h_1 \times \frac{L_1}{L_1'} \times \frac{L_2}{L_2'} = 1 \times \frac{(-8)}{(-10.5)} \times \frac{(-7.78)}{(-12.78)} = 0.464cm$.

The positive sign for h_2' indicates the image is erect. It is, of course, diminished.

---

11. Figure 9.4 shows the optical centre, O, for various thick lenses. Show that

$$A_1O = t \cdot \frac{F_2}{F_1 + F_2} \quad ,$$

where t is the axial thickness of the lens, and F_1 and F_2 are the surface powers.

Hint: use $\frac{n'}{\ell'} - \frac{n}{\ell} = F$ at the first surface and note that O and P are conjugate points.

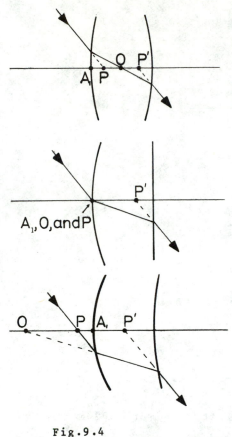

A few points are worth noting before proceeding with the proof. The incident and emergent rays are parallel. Thus, the incident ray must be directed towards the first nodal point and the emergent ray must appear to come from the second nodal point. Of course, since $n_1 = n_2'$, P and N coincide, as do P' and N'. For such an undeviated ray O is the point where that part of the ray inside the lens actually crosses the principal axis or, when projected, cuts the axis. In the former case O is said to be real, and in the latter it is virtual.

Consider the biconvex lens. At the first surface O is a real image point and P is its virtual object. $\ell' = A_1O$ and $\ell = A_1P = e = \frac{t}{n_g} \cdot \frac{F_2}{F_E}$, where n_g is the refractive index of the glass. If F_1 is the power of the front surface,

$$\frac{n_1'}{\ell'} - \frac{1}{\ell} = F_1 \quad,$$

and substituting A_1O for ℓ' and $\frac{t}{n_g} \cdot \frac{F_2}{F_E}$ for ℓ,

$$\frac{n_g}{A_1O} - \frac{1}{\frac{t}{n_g} \cdot \frac{F_2}{F_E}} = F_1$$

Fig.9.4

Making A_1O the subject,

$$\frac{n_g}{A_1O} = F_1 + \frac{1}{\frac{t}{n_g} \cdot \frac{F_2}{F_E}} = F_1 + \frac{n_g\,F_E}{t\,F_2} = \frac{t F_1 F_2 + n_g\,F_E}{t\,F_2}$$

which, on taking the reciprocal of the L.H.S. and the R.H.S. and substituting $F_1 + F_2 - \frac{t}{n_g} F_1 F_2$ for F_E, gives

$$\frac{A_1O}{n_g} = \frac{t F_2}{t F_1 F_2 + n_g F_E} = \frac{t F_2}{t F_1 F_2 + n_g (F_1 + F_2 - t F_1 F_2 / n_g)}$$

$$= \frac{t F_2}{t F_1 F_2 + n_g (F_1 + F_2) - t F_1 F_2} = \frac{t F_2}{n_g (F_1 + F_2)}$$

which gives, on multiplying both sides by n_g,

$$A_1 O = t\left(\frac{F_2}{F_1 + F_2}\right) \quad .$$

Notes. (i) The term in brackets has no units, so whatever units are given to t A_1O will have those units.
(ii) In practice we talk loosely about 'marking the optical centre'. It is evident that we generally mark A_1. The only case where we really do mark the optical centre is when $F_2 = 0$.

---

12. What kind of glass lens immersed in air will have a power independent of its thickness t?

(i) Generally, unless otherwise stated, power is taken to mean equivalent power.

Now, $F_E = F_1 + F_2 - \frac{t}{n_g} F_1 F_2$, and the term in t will disappear when F_1 or F_2 is zero. Suppose F_2 is zero, then $F_E = F_1$. Similarly, when F_1 is zero $F_E = F_2$. Hence, when one surface is plane the lens will have an equivalent power independent of t.

(ii) Consider a plano-convex lens where $F_2 = 0$. The F.V.P. will be independent of t:

i.e. $F_v = \dfrac{F_E}{1 - \frac{t}{n_g}F_2} = \dfrac{F_1}{1} = F_1$,

but the B.V.P. will be $F_v' = \dfrac{F_E}{1 - \frac{t}{n_g}F_1} = \dfrac{F_1}{1 - \frac{t}{n_g}F_1}$,

which is not independent of t. The converse is true for a plano-concave lens where $F_1 = 0$.

---

13. Suppose an object is located in the first principal plane
of a thick meniscus lens. Determine the position of the
image and the magnification.

Using Newton's equation $xx' = f_E f'_E$. Then, since $x = FP = -f_E$,
we have $x' = \dfrac{f_E f'_E}{x} = \dfrac{f f'_E}{-f_E} = -f'_E = -P'F' = F'P'$.

That is, the image is in the second principal plane.
The magnification is

$$m = -\frac{f_E}{x} = -\frac{f_E}{-f_E} = 1 .$$

That is, we have unit magnification. This is why the
principal planes are occasionally referred to as the unit
planes.

---

14. Figure 9.5a shows the principal planes and second focal
plane of a thick lens. Determine graphically the conjugate
image point of the object point B .

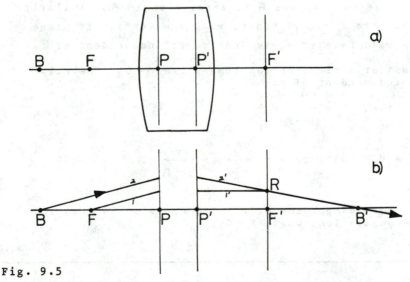

Fig. 9.5

Construct any ray 1 through F. This will leave the second
principal plane, as 1', at the same height that ray 1 met
the first principal plane, and it will be parallel to the
axis. A ray 2, through B and parallel to 1, will leave the
second principal plane as 2' at the same height as 2 met the
first plane. 2' will meet the second focal plane in the same
point as does 1' ; i.e. at R. Producing 2' to meet the axis
locates the image point, B'.

15. Figure 9.6 is a Huygens eyepiece. The first lens is known as the field lens and the second as the eyelens. If their focal lengths are f_1' and f_2', where $f_1' = 3f_2'$, and their separation is $(f_1' + f_2')/2 = 2f_2'$, find the positions of P, P', F, and F'. Note: the condition that the separation is $(f_1' + f_2')/2$ corrects for lateral chromatic aberration.

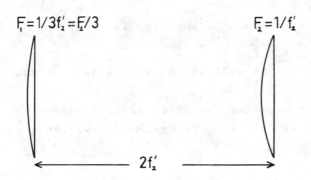

$F_1 = 1/3f_2' = F_2/3$ $F_2 = 1/f_2'$

$2f_2'$

Fig. 9.6

The equivalent power is

$$F_E = F_1 + F_2 - dF_1F_2 = \frac{F_2}{3} + F_2 - 2f_2'\frac{F_2}{3}F_2 = \frac{2}{3}F_2 \ .$$

So,

$$f_E' = \frac{1}{F_E} = \frac{1}{\frac{2}{3}F_2} = \frac{3}{2}f_2' \ .$$

and

$$f_E = -f_E' = -\frac{3}{2}f_2' \ .$$

The F.V.P. is

$$F_v = \frac{F_E}{1 - dF_2} = \frac{\frac{2}{3}F_2}{1 - 2f_2'F_2} = \frac{\frac{2}{3}F_2}{1 - 2} = -\frac{2}{3}F_2 \ .$$

Note that although F_E ($= \frac{2}{3}F_2$) is positive, F_v is negative.

Since, $F_v = -\frac{1}{f_v}$, $f_v = -\frac{1}{F_v} = -\frac{1}{(-\frac{2}{3}F_2)} = \frac{3}{2}f_2'$.

Therefore, the first focal point, F, is $A_1F = f_v = \frac{3}{2}f_2'$ to the right of the first lens.

The first principal point's position is given by

$$A_1P = f_v - f_E = \frac{3}{2}f_2' - \left(-\frac{3}{2}f_2'\right) = 3f_2' \ .$$

That is, it is f_2' to the right of the second lens.

-72-

The B.V.P. is

$$F'_v = \frac{F_E}{1-dF_1} = \frac{\frac{2}{3}F_2}{1 - 2f'_2\frac{F_2}{3}} = \frac{\frac{2}{3}F_2}{\frac{1}{3}} = 2F_2 \quad .$$

Hence, $f'_v = \frac{1}{F'_v} = \frac{1}{2F_2} = \frac{f'_2}{2}$, placing the second focal point

$f'_2/2$ to the right of the second lens.

The second principal point's position is given by

$$A_2P' = f'_v - f'_E = \frac{f'_2}{2} - \frac{3}{2}f'_2 = -f'_2 \quad .$$

That is, P' is f'_2 to the left of the second lens, midway between the two lenses.

Since F'_v (=$2F_2$) is positive, parallel light incident on F_1 is converged by the system. Figure 9.7 shows these points.

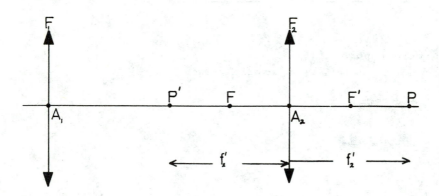

Fig. 9.7

Figure 9.8a shows parallel incident light on F_1 . If F_1 = +10D and F_2 = +30D, the vergence leaving the system is +60D. Figure 9.8b shows parallel light incident on F_2. The vergence leaving the system, for the above values of F_1 and F_2, is -20D.

Reversing these rays in 9.8b, light directed towards the first focal point of the system emerges parallel. This is the situation in an instrument where the eye is the detector. Clearly, the objective lens in the total system is responsible for the convergent light arriving at the field lens.

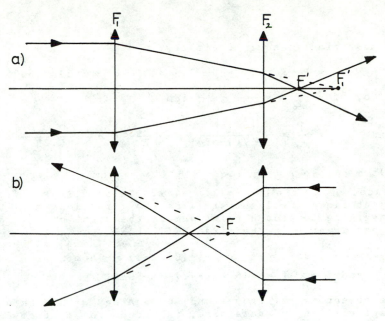

Fig. 9.8

---

16. A thin converging lens of power $3\frac{1}{3}$D is 25cm in front of a
    thin diverging lens of power -20D. What is the equivalent
    power of the system?

    We have $F_1 = +3\frac{1}{3}$D, $F_2 = -20$D, and $d = 0.25$ m

    Then, $F_E = F_1 + F_2 - dF_1F_2 = 3\frac{1}{3} + (-20) - 0.25 \times 3\frac{1}{3} \times (-20)$

    $= -16\frac{2}{3} - (-16\frac{2}{3}) = 0$.

    The system is afocal. It is a Galilean telescope in its
    infinity or afocal setting.

---

17. A thin convex lens of 20cm focal length lies 10cm in front
    of a thin convex lens of 40cm focal length. Assuming the
    system lies in air, where must an object be placed in relation
    to the first lens in order to produce an image of the same
    size and (a) erect (b) inverted?

    The magnification in each case is (a) +1, and (b) -1 . Newton's
    relationship states that $m = -f_E/x$, so that if we find x in each
    case we will know the object position relative to the first
    focal point, F, of the system. Then, if we find f_v we will know
    how far F is from the first lens. This allows us to say
    where the object is in relation to the first lens since
    $\ell_1 = x + f_v$. ℓ_1 is the object distance from the first lens.

We must find f_E and f_v. Firstly,

$$F_E = F_1 + F_2 - dF_1F_2 = 5 + 2.5 - (0.1 \times 5 \times 2.5) = +6.25D \quad .$$

So, $f_E = -\dfrac{1}{F_E} = -\dfrac{1}{6.25} \, m \equiv -16 \, cm.$

and $f_v = -\dfrac{1}{F_v} = -\dfrac{1}{F_E/(1-dF_2)} = -\dfrac{1-dF_2}{F_E} = -\dfrac{1-(0.1 \times 2.5)}{6.25} \equiv -12 \, cm.$

(a) $m = 1$, so $x = -f_E/m = -(-16)/1 = +16 \, cm.$

Hence, $\ell_1 = x + f_v = 16 + (-12) = +4 \, cm.$ This will be a virtual object since the light striking the first lens will be converging towards a point 4cm to the right of it. Note: the object is at P in the first UNIT PLANE.

(b) $m = -1$, so $x = -f_E/m = -(-16)/(-1) = -16 \, cm,$

and $\ell_1 = x + f_v = -16 + (-12) = -28 \, cm.$

The object is $-2f_E'$ from P and, by analogy with a thin lens, the image will be the same size, inverted, and $2f_E'$ from P'.

---

18. An equiconvex lens made of glass of refractive index $n_g = 1.5$ with axial thickness 6cm has surface curvatures 10D. It is bounded by air on one side and water, $n_w = 1.33$, on the other. Find the positions of the cardinal points.

We shall find F', F, P', P, N', N, in that order.

Surface powers
Refer to figure 9.9 for the data.

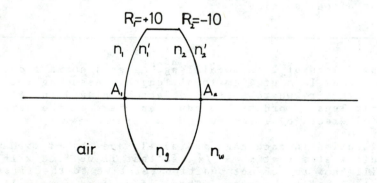

Fig. 9.9

$$F_1 = (n_1' - n_1)R_1 = (1.5 - 1) \times 10 = +5D$$
and $$F_2 = (n_2' - n_2)R_2 = (1.33 - 1.5)(-10) = +1.7D \quad .$$

Equivalent power

$$F_E = F_1 + F_2 - \frac{t}{n_g} F_1 F_2 = 5 + 1.7 - \frac{0.06}{1.5} \times 5 \times 1.7 = +6.36D$$

Back vertex power and back vertex focal length

$$F_v' = \frac{F_E}{1 - \frac{t}{n_g} F_1} = \frac{6.36}{1 - \frac{0.06}{1.5} \times 5} = +7.95D$$

Whence, $f_v' = \frac{n_2'}{F_v'} = \frac{1.33}{7.95}m \equiv 16.73cm.$

Front vertex power and front vertex focal length

$$F_v = \frac{F_E}{1 - \frac{t}{n_g} F_2} = \frac{6.36}{1 - \frac{0.06}{1.5} \times 1.7} = +6.82D$$

and $f_v = -\frac{n_1}{F_v} = -\frac{1}{6.82}m \equiv -14.66cm.$

Positions of F and F'

$$A_1F = f_v = -14.66cm$$

and $\quad A_2F' = f_v' = +16.73cm.$

Positions of P and P'

$$A_1P = e = n_1 \frac{t}{n_g} \cdot \frac{F_2}{F_E} = 1 \times \frac{6}{1.5} \times \frac{1.7}{6.36} = +1.07cm;$$

i.e. P is 1.07cm to the right of the first vertex.

$$A_2P' = e' = -n_2' \frac{t}{n_g} \cdot \frac{F_1}{F_E} = -1.33 \times \frac{6}{1.5} \times \frac{5}{6.36} = -4.18cm.$$

i.e. P' is 4.18cm to the left of the second vertex.

Positions of N and N'

$$FN = P'F' = f_E' = \frac{n_2'}{F_E} = \frac{1.33}{6.36} = +0.2091m \equiv +20.91cm;$$

i.e. N is 20.91cm to the right of F.

$$FN' = PF = f_E = -\frac{n_1}{F_E} = -\frac{1}{6.36}m \equiv -15.72cm;$$

i.e. N' is 15.72cm to the left of F'.

19. Two identical thin convex lenses are fixed in a short tube 0.4 times the focal length of either lens apart. They are mounted on an optical bench where they produce a real image magnified 2 times. When the tube is moved 25cm further from the source the magnification is 0.4. Find (a) the focal length of the combination, and (b) the focal length of each lens.

We shall use Newton's relationship for magnification where f'_E is the focal length of the combination. Let x'_1 be the extrafocal distance in the first case and x'_2 be the extra-focal distance in the second case. Since the system is moved 25cm to the right in the second case, $x'_2 = (x'_1 - 25)$ cm.

The magnification in the two cases is

$$m_1 = -2 = -\frac{x'_1}{f'_E}$$

or, $\quad x'_1 = 2 f'_E$ ....(1)

and $\quad\quad\quad m_2 = -0.4 = -\frac{x'_2}{f'_E}$

or, $\quad x'_2 = 0.4 f'_E$ ....(2)

Hence, $\quad x'_2 = (x'_1 - 25) = 2 f'_E - 25 = 0.4 f'_E$

which gives $1.6 f'_E = 25 \quad$ or, $\quad f'_E = 25/1.6 = +15.625$ cm.

Now, $\quad F_E = \frac{1}{f'_E} = \frac{1}{+0.15625} = +6.4D$

and $\quad F_E = F_1 + F_2 - d F_1 F_2$ ....(3)

Putting $F_1 = F_2 = F$, since the thin lenses are identical,

and $d = 0.4 f' = 0.4 \frac{1}{F}$, where $f' = \frac{1}{F}$, in equation (3),

$$F_E = 6.4 = F + F - \frac{0.4}{F} F.F = 2F - 0.4F = 1.6F .$$

So, $\quad F = \frac{6.4}{1.6} = +4D$.

That is, the (equivalent) power of the combination is +6.4D and each thin lens is +4D.

10. OPTICAL SYSTEMS

1. An object 1cm high is placed 15cm from a positive lens, $f' = +20$cm; what will be the visual angle subtended by the image at an eye 10cm from the lens?

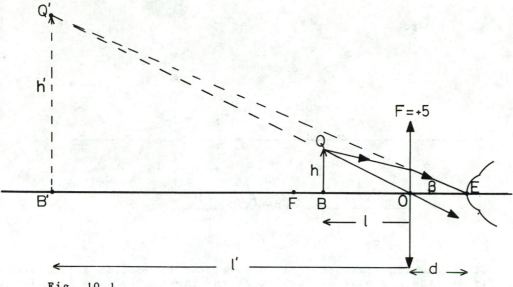

Fig. 10.1

The angular subent of the image at the eye is β. For small β, β rad $\simeq \tan\beta$, so $\beta \simeq h'/(-\ell' + d)$. We are given h, f', and ℓ so we can find ℓ' and h'.

Making ℓ' the subject of the thin lens equation $\frac{1}{\ell'} - \frac{1}{\ell} = \frac{1}{f'}$,

we have

$$\ell' = \frac{\ell \, f'}{\ell + f'} = \frac{(-15) \times 20}{(-15) + 20} = -60\text{cm}.$$

The magnification equation gives

$$h' = h.\frac{\ell'}{\ell} = 1 \times \frac{(-60)}{(-15)} = +4\text{cm}.$$

Hence, $\beta = h'/(d-\ell') = 4/(10-(-60)) = 0.05714$ rad $= 3.27^0$.

2. What must be the position of an object in order that an eye
   10cm behind a +10D lens may see it clearly when it is accommodated
   for a distance of 50cm? If the lens has a diameter of 2cm what
   length of object can be seen?

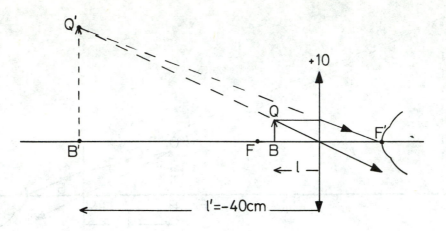

Fig. 10.2

Since the lens is 10cm from the eye, and the ocular accommodation
is 2D, the eye must be looking at the image 50cm from it which is
in turn, 40cm from the lens; see figure 10.2 . We can find the
object position from the thin lens equation:

$$\frac{1}{\ell} = \frac{1}{\ell'} - \frac{1}{f'} = \frac{1}{-40} - \frac{1}{+10} = -\frac{5}{40} , \qquad \text{so } \ell = -\frac{40}{5} = -8\text{cm} .$$

To find the maximum size object we must assume that the eye pupil
is the aperture stop. The ray from the top of the object to the
centre of the aperture stop (the cornea, here!) is the chief ray
from the top of the object. If this ray meets the uppermost edge
of the lens this will determine the maximum image size.

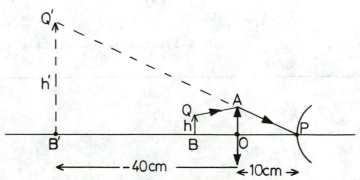

Fig. 10.3

We can find h', figure 10.3, from triangles POA and PB'Q'.
Thus, from the geometry,

$$h' = B'P \cdot \frac{AO}{OP} = 50 \cdot \frac{1}{10} = +5cm, \text{ since } AO = \tfrac{1}{2} \times 2 = 1cm.$$

We can now find h using the magnification equation,

$$h = h' \cdot \frac{\ell}{\ell'} = 5 \times \frac{(-8)}{(-40)} = +1cm.$$

Since the object can be placed on either side of the axis, the
object can be 2cm across. In the case considered here the ray
QA will be parallel to the axis since the object and the half-
diameter of the lens are each 1cm.

---

3. A microscope objective has a focal length of 4cm and the eyepiece's
focal length is 5cm. If the distance between the lenses is 20cm
find the magnification of the instrument.

Figure 10.4 shows the set-up.

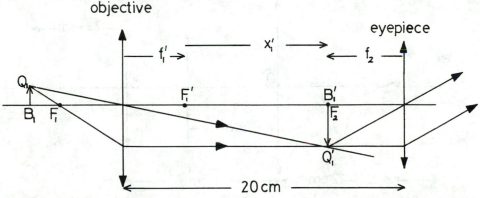

Fig. 10.4

The total magnification is composed of two factors: m_1, the
lateral magnification due to the objective, and M_2, the angular
magnification due to the eyepiece.

Hence, the total (angular) magnification is

$$m_1 \times M_2 = -\frac{x_1'}{f_1'} \times \frac{-q}{f_2'} = -\frac{11}{4} \times \frac{-(-25)}{5} = -13.75 ,$$

where $x_1' = 20 - (f_1' + f_2')$, $f_1' = +4cm$, $f_2' = +5cm$, and $q = -25cm$.

---

4. Define aperture stop, entrance pupil, and exit pupil. Two
thin positive lenses are separated by 5cm. Their diameters
are 6cm and 4cm respectively, and their focal lengths are
$f_1' = 9$cm and $f_2' = 3$cm. A diaphragm 1cm in diameter is placed
between them, 2cm from the second lens. Find the aperture
stop and the entrance and exit pupils for an axial object
point 12cm in front of the first lens.
*This question is repeated with a somewhat different treat-
ment as question 17.1 .

Definitions

Aperture stop: this is the element, a diaphragm or the rim
of a lens, which determines the amount of light reaching the
image.
Entrance pupil: this is the image of the aperture stop seen
from an axial object point.
Exit pupil: this is the image of the aperture stop seen from
an axial image point.
In a case where there are no refracting elements in front of
the aperture stop, the latter is also the entrance pupil. A
similar situation holds when there are no refracting elements
behind the aperture stop.

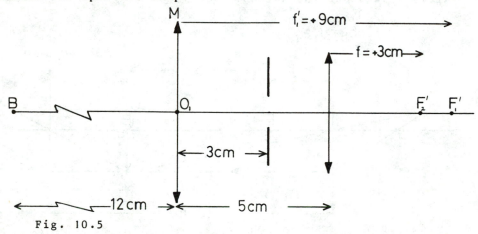

Fig. 10.5

To determine which of the three components is the aperture
stop we must find which component, or its image in any
preceding components, subtends the smallest angle at the
axial object point B.

First lens

This subtends an angle $\arctan\frac{MO_1}{BO_1} = \arctan\frac{3}{12} = 14^0$.

The image of the diaphragm in the first lens

The diaphragm is 3cm from the first lens which has a focal
length of +9cm. For convenience, imagine the object (the
diaphragm) and the lens as in figure 10.6

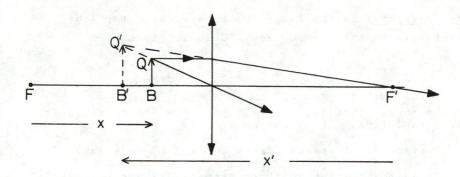

Fig. 10.6

Using Newton's relationship $xx' = -f'^2$, where $x = +6$cm and $f' = +9$cm, we have

$$x' = -\frac{f'^2}{x} = -\frac{9 \times 9}{6} = -13.5\text{cm} \ .$$

That places the image of the diaphragm 4.5cm from the first lens and 0.5cm from the second lens.

The radius of the image of the diaphragm is
$$B'Q' = BQ.(-\frac{f}{x}) = 0.5 \times \frac{(-(-9))}{6} = 0.75\text{cm}.$$

The angular subtent of this radius at the object point B, 12cm from the first lens is

$$\arctan\frac{0.75}{16.5} = 2.6^0$$

The image of the second lens in the first lens

Precisely the same calculation as before but with $x = +4$cm yields
$$x' = -\frac{f'^2}{x} = -\frac{81}{4} = -20.25\text{cm} \ ;$$

i.e. the image is 11.25cm to the right of the first lens. The radius of the image is given by

$$\text{radius of second lens} \times (-\frac{f}{x}) = 2 \times (-(\frac{-9}{4})) = 4.5\text{cm}.$$

This image is $11.25 + 12 = 23.25$cm from the axial object point and subtends an angle of $\arctan(4.5/23.25) = 10.95^0$.

The smallest of these three angles determines the aperture stop. Accordingly, the diaphragm is the aperture stop, and its image in the first lens is the entrance pupil.

The <u>exit pupil</u> is the image of the aperture stop (the diaphragm) in the second lens. This is given by

$$\frac{1}{\ell'} = \frac{1}{\ell} + \frac{1}{f'} = \frac{1}{-2} + \frac{1}{3} = \frac{-3 + 2}{6} = -\frac{1}{6} \quad ,$$

so $\ell' = -6$cm. That is, the exit pupil is 6cm to the left of the second lens. Its diameter is

diameter of aperture stop $\times \dfrac{\ell'}{\ell} = 1 \times \dfrac{(-6)}{(-2)} = 3$cm.

Figure 10.7 shows the aperture stop (A.S.) and the pupils, and a ray directed towards the centre of the entrance pupil leaving the system as though from the centre of the exit pupil.

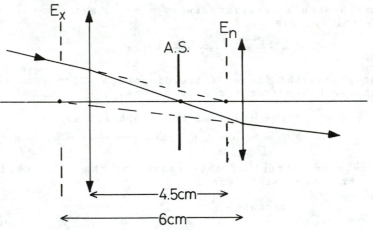

Fig. 10.7

5. A telescope consists of a +12.5D eyepiece and a +2.5D objective. What is the magnifying power of the telescope (a) in its infinity (afocal) adjustment and (b) when the final image is viewed by the observer at a distance of 50cm? Assume the eye is close to the eyepiece.

(a) The magnifying power of a telescope in its infinity setting is defined as the ratio of the angular subtent of the image seen with the instrument to the angular subtent of the object seen without the instrument. That is, in figure 10.8, the angular magnification is

$$M = \beta/\alpha$$

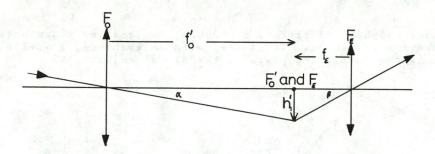

Fig. 10.8

Now, $\alpha = -h_i'/f_o'$ and $\beta = h_i'/f_\varepsilon$, where f_o' is the focal length of the objective and f_ε' is the focal length of the eyepiece. Note that α is positive, β is negative, and $f_\varepsilon' = -f_\varepsilon$.

Hence, $M = \dfrac{\beta}{\alpha} = \dfrac{h_i'}{f_\varepsilon} \Big/ -\dfrac{h_i'}{f_o'} = -\dfrac{f_o'}{f_\varepsilon'} = -\dfrac{40}{8}\dfrac{cm}{cm} = -5$

We have made use of the fact that $f' = 1/F$ for a thin lens.

(b) When the final image is 50cm from the eye h_i' must be inside the first focal plane of the eyepiece, figure 10.9 .

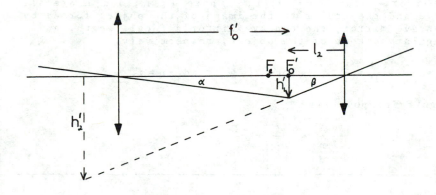

Fig. 10.9

β is now $-\dfrac{h_i'}{\ell_2} = -h_i' \Big/ \dfrac{\ell_2\, f_\varepsilon'}{f_\varepsilon' - \ell_2} = -h_i' \Big/ \left(\dfrac{(-50)\times 8}{8 - (-50)}\right) = 0.145 h_i'$.

Thus, $M = \dfrac{\beta}{\alpha} = \dfrac{0.145 h_i'}{-h_i'/f_o'} = -0.145 f_o' = -0.145 \times 40 = -5.8$.

6. A concave spherical mirror, $r = -200$cm, is used for a Newtonian reflecting telescope. It is used with an eyepiece, $F_\epsilon = +25$D. What is the magnifying power in the afocal adjustment?

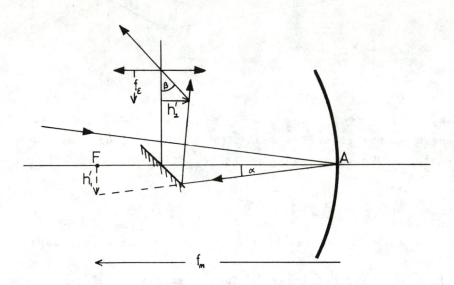

Fig. 10.10

h_1' would be the image formed in the absence of the secondary plane mirror. Hence, $\alpha = -h_1'/f_m$; note h_1', f_m, and α are all negative and they refer to the image of the object formed by the concave mirror, the concave mirror's focal length, and the angular subtent of the object, respectively.

$$\beta = -\frac{h_2'}{f_\epsilon} = \frac{h_2'}{f_\epsilon'} = \frac{h_1'}{f_\epsilon'} \quad , \quad \text{since } h_1' = h_2' \text{ , and } -f_\epsilon = f_\epsilon' \ .$$

The magnifying power is

$$M = \frac{\beta}{\alpha} = \frac{h_1'/f_\epsilon'}{-h_1'/f_m} = -\frac{f_m}{f_\epsilon'} = -\frac{(-100)}{4} = 25$$

---

7. Define the f/Nº (or f/#) used on a camera lens. What are the common f-numbers chosen and what is the reason for choosing them?

The f-number is defined as the ratio of the camera lens focal length, f', to the aperture stop diameter, D; i.e. f/# = f'/D .

The most common series chosen is 2, 2.8, 4, 5.6, 8, 11, 16, 22. Since the focal length, f', is constant, doubling the f/#

corresponds to halving the diameter of the aperture stop.
The reason for this choice of apertures is as follows. The
luminous flux passing through the aperture is proportional
to its area; i.e. $\propto D^2$ since the area is $\pi(D/2)^2$.
But $D^2 \propto 1/(f/\#)^2$, so the intensity of light at the film
plane, I, is proportional to $1/(f/\#)^2$, or $I \propto 1/(f/\#)^2$. To
halve the intensity, I, we must increase the f-number by $\sqrt{2}$.
Notice that in the series of f-numbers given above each
increases by a factor of approximately $\sqrt{2}$ from left to right.
This means each time we 'stop down one', say from 8 to 11, and
at the same time keep the exposure time constant, the intensity
at the film plane is halved.

A useful quantity relating exposure time and intensity may
be defined as follows:

$$\text{exposure, } E = I.\Delta t \propto \frac{\Delta t}{(f/\#)^2}$$, where Δt is the exposure

time. Clearly, to keep the exposure constant when we increase
the f-number by $\sqrt{2}$ we must double the exposure time. For
example, suppose we have an exposure with f/5.6 and $\Delta t = 1/120$ s,
then if we change the aperture to f/8 we must change the
exposure time to 1/60 s. Note that $8 \simeq 5.6 \times \sqrt{2}$. Alternatively,
if we change from f/5.6 and $\Delta t = 1/120$ s to f/4, and we are to
keep the exposure the same, Δt must be reduced by approximately
half to 1/250 s.

. If a photograph of a moving merry-go-round is perfectly
exposed, but blurred, at 1/30s and f/11, what must be the
diaphragm setting if the shutter speed is raised to 1/120s
in order to 'stop' the motion?

We are told that the exposure E is correct.

$$\text{Now, } E \propto \frac{\Delta t}{(f/\#)^2} = \frac{1/30}{11^2}$$

and E must remain the same when $\Delta t = 1/120$s.

Thus, $$\frac{1/30}{11^2} = \frac{1/120}{(f/\#)^2} ,$$

so the new f-number is

$$f/\# = \left(\frac{\frac{1}{120} \times 11^2}{1/30}\right)^{\frac{1}{2}} = 5.5 .$$

The nearest stop on a camera will be the f/5.6 .

---

11. PHOTOMETRY

1. Discuss the meaning of the terms luminous intensity, luminous flux, illuminance (illumination), and luminance. A street lamp illuminates a point on the horizontal road surface with an illuminance of 0.1 lm m^{-2} (lux). The lamp is 6m above the street and 6m horizontally to one side of the point. What is the lamp's intensity?

In loose terms, any lighting system and the illuminated environment requires four defined terms. The following quantities must be defined:

    (i) the 'strength' of the source
   (ii) the rate at which light energy flows from the source in a specified direction,
  (iii) the amount of light falling on a unit area of surface perpendicular to the direction in which the light is travelling,
   (iv) the amount of light reflected back from a unit area of surface in a given direction.

Luminous intensity refers to the strength of a source in a given direction. It is a measure of the rate of flow of light energy per unit solid angle in the given direction and is measured in candela (cd). The luminous intensity is given by $I = F/w$, where F is the rate of flow of light and w is the solid angle into which it flows.

Luminous flux is the rate of flow of light energy from a source. Since it is a rate of flow of energy it expresses the power of the source. It is given the symbol F and is measured in lumens (lm).

Illuminance, symbol E, is the measure of the luminous flux density at a point on an illuminated surface; that is, it is the amount of light flowing each second through a unit area. It is measured in lumens per unit area, e.g. lumen/square metre (or lux) and is therefore a 'power density'.

Luminance, symbol L, is the amount of light per unit area emitted or re-emitted by a surface in a given direction. Where it refers to reflected light from a surface it is related to illuminance by the equation

   luminance = illuminance × reflectance (reflection coefficient).

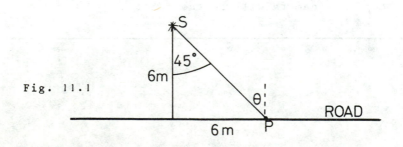

Fig. 11.1

The illuminance at the point P in figure 11.1 is given by

$$E = \frac{I}{SP^2} \cos\theta \quad .$$

So, the intensity of the source, assumed uniform in all directions, is

$$I = \frac{E \cdot SP^2}{\cos\theta} = \frac{0.1 \times (6/\cos 45^0)^2}{\cos 45^0}$$

$$= \frac{0.1 \times 36}{\cos^3 45^0} = \frac{3.6}{(1/\sqrt{2})^3} = 3.6 \times 2\sqrt{2}$$

$$= 10.2 \text{ cd}.$$

---

2. What is the illuminance on a screen when a 100cd point source is placed 25cm in front of a thin +9.00D lens with a circular aperture of 20mm diameter, and the screen is 15cm from the lens?

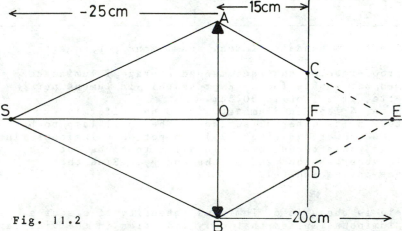

Fig. 11.2

We must find the flux occupying the solid angle formed by the area of the lens subtended at the source, S. This flux falls on the screen over the area with diameter CD. Dividing this flux by the illuminated area of the screen will give us the illuminance (illumination).

The flux is given by F = Iw, where I is the intensity of the source and w is the solid angle previously mentioned. Now,

$$w \simeq \frac{\text{area of lens}}{SO^2} = \frac{\pi AO^2}{SO^2} = \frac{\pi \times 1^2}{25^2} = \frac{\pi}{25^2} \text{ st.}$$

Since $I = 100cd$, $F = Iw = 100 \times \dfrac{\pi}{25^2}$ 1m.

To find the area of illuminated disc on the screen we must find its radius CF. From similar triangles AOE and CFE,

$$CF = AO.\frac{FE}{OE} = 1 \times \frac{5}{20} = 0.25cm = 0.0025m,$$

OE having been obtained by application of the thin lens equation,

$$\frac{1}{\ell'} - \frac{1}{\ell} = \frac{1}{f'} \quad .$$

The area of the illuminated patch on the screen is

$$\pi.CF^2 = (25 \times 10^{-4})^2 \pi \ m^2 ,$$

whence the illuminance is

$$\frac{\text{flux incident on the illuminated patch}}{\text{area of patch}}$$

$$= \frac{100 \times \pi/25^2}{(25 \times 10^{-4})^2 \pi} = 25600 \ \text{lux} .$$

---

3. Explain the terms candela, lumen, and lumens per square foot.
    (a) For proof-reading the recommended average illuminance is 20 lumen per square foot. Express this in lumens per square metre, if 1 foot is 30.5cm.
    (b) A sphere, 4 feet in diameter, is to be uniformly illuminated on its internal surface. The small lamp to be used has 100W power and gives 12 lumen per watt luminous flux. The light is radiated uniformly in all directions but the pearl bulb absorbs about 20% of the energy. Find the illuminance on the sphere.

The candela is the unit of luminous intensity of a source. The word luminous implies that only radiation in the visible spectrum is considered. The candela (cd) is defined as one sixtieth of the luminous intensity from $1cm^2$ of the surface of molten platinum at its freezing point ($1\,773^0C$) under standard atmospheric pressure ($101\,325$ newtons).

The lumen is the unit used to measure the rate of flow of light, the flux, from a source. If light from a 'point' source of intensity I flows into a solid angle of w steradians, then the flux is given by $F = Iw$. Clearly, one lumen will be achieved when a source of 1 candela outputs its luminous energy into a solid angle of 1 steradian.

The lumen per square foot is a measure of the flux density, or illuminance. It is a flux of 1 lumen falling on a surface area of 1 square foot, the surface being perpendicular to the direction of propagation.

(a) Since 20 lumens fall on 1 square foot, we can write
this as 20 lm falling on 0.305^2 m². That is, the illuminance,
E, is

$$E = \frac{20}{0.305^2} = 20 \times 10.75 = 215 \text{ lm m}^{-2}.$$

(b) The effective output of the lamp is 80% of 1200 lm = 960 lm,
and this flows into 4π steradians.
The intensity of the lamp is $I = F/w = 960/4\pi$ cd. , whence
the illuminance is $E = I/d^2 = (960/4\pi)/2^2$ lm per square foot.
d = 2ft, the radius of the sphere, and the lamp is at the
sphere's centre, of course.

The uniform illuminance on the sphere is therefore

$(960/4\pi)/2^2 = 19.1$ lm per square foot.

---

4.   A small lamp, emitting luminous flux uniformly in all
directions, is situated 4ft above the centre of a circular
table of diameter 6ft. If the intensity of illumination
at the table's edge is 3.2 lm per square foot, calculate
        a) the intensity of the lamp
        b) the illuminance at the centre of the table
        c) the total flux falling on the table
  and d) the mean illuminance on the table.

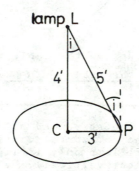

Fig. 11.3

Figure 11.3 shows the lamp, L, the centre, C, and a point, P,
on the table. The triangle LCP is clearly a 3,4,5 right
angled triangle.

a) The illuminance at P is $E_P = \frac{I}{LP^2}\cos i$, so the intensity
of the lamp is

$$I = \frac{E_P \cdot LP^2}{\cos i} = \frac{3.2 \times 5^2}{4/5} = 100 \text{ cd}.$$

b) The illuminance at C is

$$E_c = \frac{I}{LC^2} = \frac{100}{4^2} = 6.25 \text{ lm per sq. ft.}$$

c) The total flux emitted by the lamp into 4π steradians is $4\pi I = 4\pi \times 100$ lm. The table subtends approximately $9\pi/25$ ster. at the lamp, given by

$$\frac{\text{area of table}}{LP^2} = \frac{\pi.CP^2}{LP^2} = \frac{3^2\pi}{5^2} = \frac{9\pi}{25} \text{ ster.}$$

If we express the angular subtent of the table as a fraction of 4π steradians this gives that fraction of the total flux which falls on the table. Since the total flux is $F = Iw = 100 \times 4\pi$ lm, the flux falling on the table is

$$\frac{9\pi/25}{4\pi} \times (100 \times 4\pi) = 36\pi \text{ lm.}$$

The radius of the sphere into which the lamp radiates is 5 feet, and the table's area has been taken as approximately equal to that area of the sphere bounded by the solid angle, figure 11.4 .

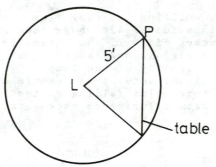

Fig. 11.4

d) The mean illuminance on the table is given by

$$\frac{\text{flux falling on the table}}{\text{area of table}} = \frac{36\pi}{\pi \times 3^2} = 4 \text{ lm per sq. foot.}$$

---

5. Two identical lamps are suspended at a height h above a horizontal floor; the horizontal distance between the lamps is 12 feet. When one lamp only is lighted the illuminance at a point on the floor vertically below the lamp is 5 lm ft^{-2}; when both lamps are on it is 6.08 lm ft^{-2}. Find h and the intensity of each lamp.

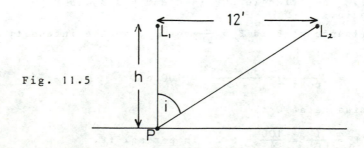

Fig. 11.5

Figure 11.5 shows the lamps, L_1 and L_2, and the point P below the lamp L_1. The illuminance at P due to the lamp L_1 is

$$E_1 = \frac{I}{h^2} = 5 \text{ lm ft}^{-2} \text{ , where I is the lamp's intensity...(1)}$$

The illuminance at P due to L_2 is $6.08 - 5 = 1.08 \text{ lm ft}^{-2}$, and this is given by

$$E_2 = 1.08 = \frac{I}{L_2P^2} \cos i = \frac{Ih}{L_2P^3} \text{ , since } \cos i = \frac{h}{L_2P} \quad \ldots\ldots\ldots(2)$$

Since $L_2P = (h^2 + 12^2)^{\frac{1}{2}}$, equation (2) becomes

$$1.08 = \frac{Ih}{\sqrt{(h^2 + 12^2)^3}} \quad \ldots\ldots\ldots\ldots\ldots\ldots\ldots\ldots\ldots(3)$$

Substituting for I from equation (1) into equation (3),

$$1.08 = \frac{5h^3}{\sqrt{(h^2 + 12^2)^3}} \text{ ;}$$

i.e. $1.08\sqrt{(h^2 + 12^2)^3} = 5h^3$.

Squaring both sides,

$$1.08^2 (h^2 + 12^2)^3 = 25h^6 .$$

Taking cube roots of both sides,

$$1.08^{\frac{2}{3}}(h^2 + 12^2) = 25^{\frac{1}{3}}h^2$$

or, $h = 8\frac{2}{3}$ ft , taking the positive root after some arithmetic.

Substituting for h in equation (1),

$$I = 5h^2 = 5 \times 8.67^2 = 5 \times 75 = 375 \text{ cd.}$$

---

6. A laser emits a 2.5mm diameter beam of collimated light with a radiant flux of 200mW. Calculate its irradiance. How would you set-up a -40D lens and a convex lens to expand the beam so that its irradiance is reduced by a factor of 4?

The cross-sectional area of the beam is $\pi(1.25 \times 10^{-3})^2 \text{ m}^2$. Hence, the irradiance is

$$\frac{\text{power}}{\text{area}} = \frac{200 \times 10^{-3}}{\pi(1.25 \times 10^{-3})^2} \quad \frac{W}{m^2}$$

$$= 40.7 \times 10^3 \text{ Wm}^{-2} .$$

Note: if the laser is emitting in the visible spectrum we may convert to illuminance using the conversion factor $1 \text{ W} \equiv 679.6 \text{ lm}$. The lumen is a relative unit of power based on the candela.

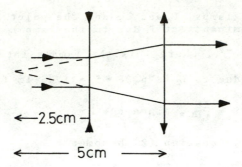

←2.5cm→

←——— 5cm ———→

Fig. 11.6

Figure 11.6 shows a beam expander set-up. Clearly, it is
a reversed Galilean telescope, in principle. Since we want
to reduce the irradiance to one quarter its value with a 2.5mm
diameter beam, we must quadruple the beam cross-sectional
area. This simply means we must double its radius.

From similar triangles in figure 11.6 it should be evident
that the diameter of the beam will double if the convex lens
has a focal length twice that of the minus lens, and the
separation is equal to the focal length of the minus lens.
Hence, the power of the convex lens must be +20D and the
separation 2.5cm.

───

12. THE WAVE NATURE OF LIGHT

1. Light of wavelength $\lambda = 589$nm, in vacuo, is associated with
    a refractive index of 2.417 for diamond and 1.923 for zircon.
    Calculate the ratio of its wavelength in diamond to that in
    zircon.

    The wavelength in diamond is $\lambda_d = 589/2.417$ nm, and
    the wavelength in zircon is $\lambda_z = 589/1.923$ nm.

    Hence, $\lambda_d/\lambda_z = \dfrac{589}{2.417}/\dfrac{589}{1.923} = 0.7956$

---

2. What do you understand by the terms wavelength, amplitude,
    and phase? Illustrate your answer with reference to the
    functions $A \sin 2\pi \dfrac{x}{\lambda}$ and $A \sin 2\pi \dfrac{t}{T}$.

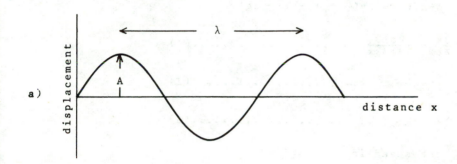

a)

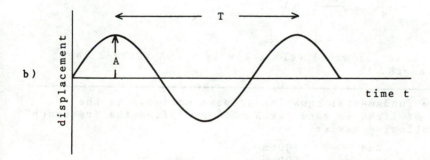

b)

Fig. 12.1

Figure 12.1a shows a plot of the function $\Psi(x) = A \sin 2\pi \dfrac{x}{\lambda}$.

Since the sine function has maximum and minimum values of
+1 and -1, $\Psi(x)$ has maximum and minimum values of +A and -A.
A is called the amplitude of the sine wave (or graph).

Note that $\Psi(x+\lambda) = A \sin 2\pi (\dfrac{x+\lambda}{\lambda}) = A \sin (2\pi\dfrac{x}{\lambda} + 2\pi)$

$$= A \sin 2\pi\dfrac{x}{\lambda} = \Psi(x) .$$

That is, $\Psi(x) = \Psi(x+\lambda)$, which indicates that the function takes the same value at intervals or periods of λ. When x and λ are distances the repeat distance λ is called the wavelength or spatial period. If t and T are times in the function $\Psi(t) = A\sin 2\pi\frac{t}{T}$, the repeat time when $\Psi(t) = \Psi(t+T)$ is T, and T is called the period or periodic time.

The quantity $2\pi\frac{x}{\lambda}$ in $A\sin 2\pi\frac{x}{\lambda}$ is called the phase.

3.* $\Psi(x,t) = A\sin 2\pi(\frac{x}{\lambda} - \frac{t}{T})$ represents a sinusoidal wave travelling in the positive x direction. If we write $k = 2\pi/\lambda$ and $\omega = 2\pi/T = 2\pi\nu$, where $\nu = 1/T$ is the frequency, find the speed of the wave's propagation $v = \frac{\partial x}{\partial t}$.

The phase is a function of two variables x and t. Calling the function ϕ, we can write

$$\phi(x,t) = 2\pi(\frac{x}{\lambda} - \frac{t}{T}) = kx - \omega t .$$

Then $\quad \frac{\partial x}{\partial t} = -\frac{\partial\phi/\partial t}{\partial\phi/\partial x}$ ....(1)

But $\quad \frac{\partial\phi}{\partial t} = -\omega$ ....(2)

and $\quad \frac{\partial\phi}{\partial x} = k$ ....(3)

Hence, the speed is

$$\frac{\partial x}{\partial t} = -(\frac{-\omega}{k}) = \frac{2\pi\nu}{2\pi/\lambda} = \nu\lambda \quad ;$$

i.e. the phase (or wave) velocity v is $v = \nu\lambda$ where λ is the wavelength.

---

4. State the fundamental equation of wave motion. If the velocity of light in vacuo is 3×10^8 ms^{-1} find the frequency of the following waves:

   (i) Red   700nm
   (ii) Orange 600nm
   (iii) Violet 400nm.

The fundamental equation of wave motion is
   speed = frequency × wavelength
or, in symbols, $v = \nu\lambda$. Given v and λ we can find the frequencies:

(i) $\nu = v/\lambda = 3 \times 10^8 / 700 \times 10^{-9} = 4.29 \times 10^{14}$ Hz,

(ii) $\nu = v/\lambda = 3 \times 10^8 / 600 \times 10^{-9} = 5.0 \times 10^{14}$ Hz,

(iii) $\nu = v/\lambda = 3 \times 10^8 / 400 \times 10^{-9} = 7.5 \times 10^{14}$ Hz.

---

5. Overhead power lines carry alternating current with a frequency of 50 Hz. Find the wavelength of the electromagnetic radiation from the wires.

$\lambda = v/\nu = 3 \times 10^8 / 50 = 6 \times 10^6$ m.

---

6. Given the travelling wave

$$\Psi(x,t) = 20 \sin 2\pi \left(\frac{x}{6 \times 10^{-7}} - \frac{t}{2 \times 10^{-15}} \right)$$

find the frequency, wavelength, and speed. Use S.I. units.

By comparison with the general form of the equation

$\Psi(x,t) = A \sin 2\pi \left(\frac{x}{\lambda} - \frac{t}{T} \right)$ we can see immediately that

$\lambda = 6 \times 10^{-7}$ m, and $\nu = 1/T = 1/(2 \times 10^{-15}) = 5 \times 10^{14}$ Hz.

Hence, the speed $v = \nu\lambda = 5 \times 10^{14} \times 6 \times 10^{-7} = 3 \times 10^8$ ms^{-1}.

---

7. A source is sending out waves in a continuous medium. The displacement from the mean position is given by the function $\Psi(x,t) = 10 \sin 2\pi \left(\frac{x}{\lambda} - \frac{t}{T} \right)$ and the waves travel at 200 cm s^{-1}. Given that $T = \frac{1}{3}$ s, find the displacement of a particle 150 cm from the source after a time of 1 s.

The frequency $\nu = 1/T = 1/\frac{1}{3} = 3$ Hz. Hence, the wavelength is $\lambda = v/\nu = 200/3$ cm, and the displacement is

$$\Psi(x,t) = 10 \sin 2\pi \left(\frac{x}{200/3} - \frac{t}{1/3} \right) = 10 \sin 6\pi \left(\frac{x}{200} - t \right)$$

$$= 10 \sin 6\pi \left(\frac{150}{200} - 1 \right), \text{ for } x = 150 \text{ cm and } t = 1 \text{ s},$$

$$= 10 \sin \left(-\frac{3}{2}\pi \right) = 10 \times 1 = 10 \text{ cm}.$$

---

8. Sketch the curves $\sin\theta$, $\sin(\theta-\frac{\pi}{2})$, $\sin(\theta-\pi)$, and $\sin(\theta-\frac{3}{2}\pi)$. Use these graphs to explain why $\sin 2\pi(\frac{x}{\lambda}-\frac{t}{T})$ represents a sine wave travelling in the positive x direction.

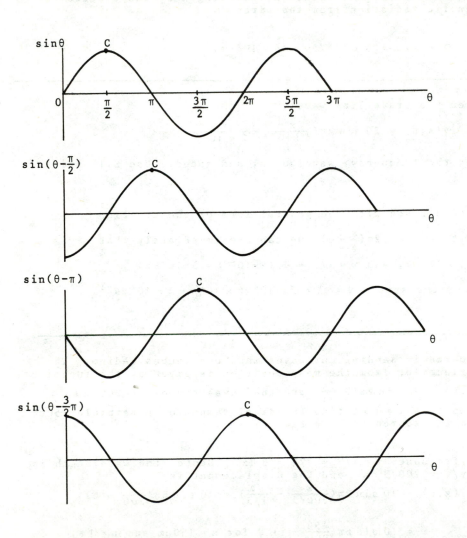

Fig. 12.2

Notice how the crest C moves successively in the positive θ direction. Now, if θ = 2πx/λ and 2πt/T is a continuous function of time, t, then as t increases the crest C will move smoothly in the positive x direction.

---

9. At time t = 0 a wave has the form $\Psi(x,0) = 3 \sin(\frac{\pi x}{20})$.

If the wave is moving in the positive x direction write the equation for the disturbance at t = 8 s, when v = 2ms^{-1}.

The wave must have the form $\Psi(x,t) = A \sin 2\pi(\frac{x}{\lambda} - \frac{t}{T})$. We can see immediately that A = 3m, and if we rewrite $\Psi(x,0)$ as

$$\Psi(x,0) = 3 \sin 2\pi(\frac{x}{40})$$

we can see that λ = 40m.

Using the relationship $v = \nu\lambda = \frac{\lambda}{T}$, we have T = λ/v = 40/2 = 20 s.

Hence, for t = 8s we have

$$\Psi(x,8) = 3 \sin 2\pi(\frac{x}{40} - \frac{8}{20}) .$$

---

10. In a Fizeau's method experiment for measuring the speed of light the toothed wheel rotates 5590 times per minute. The wheel has 100 teeth and 100 spaces and the mirror is 8.05km distant. Find the speed of light.

The light will be cut off if it returns after reflection and is met by a tooth. This will occur first if a tooth moves into the space through which the light passed in the time taken for the light to travel to the mirror and back.

Let the speed of light be c km/s. Then the time taken to travel to the mirror and back, a distance of 2 × 8.05 km, is 16.10/c s.

Since there are 200 'teeth-spaces' and the wheel rotates 5590/60 times per second, (5590/60) × 200 teeth-spaces will cross the light path per second. Hence, one tooth will take $\frac{1}{\frac{5590}{60} \times 200}$ s to move into the position previously occupied by a space. These times must be equal to cut off the light, hence,

$$\frac{16.10}{c} = \frac{1}{\frac{5590}{60} \times 200}$$

or, $c = 16.10 \times \dfrac{5590}{60} \times 200 = 2.999967 \times 10^5$ km/s.

---

11. Suppose the previous experiment were conducted in a tube
    which is now filled with water of refractive index 4/3.
    What is the number of revolutions per minute through which
    the wheel must now rotate to first cut off the light?

    The speed of light will be $3 \times 10^5/(4/3)$ km/s;
    i.e. $\frac{3}{4} \times 3 \times 10^5$ km/s.

    Replacing c in the equation of the last question by the new
    speed of light, and replacing 5590 revolutions per minute by
    the unknown R, we have, on rearranging,

    $$R = \frac{\frac{3}{4} \times 3 \times 10^5 \times 60}{16.10 \times 200} = 4192.5 \text{ revs/min.}$$

    Or, more simply, since the light travels at $\frac{3}{4}$ its speed in
    air the wheel must rotate at $\frac{3}{4}$ of its original speed;
    i.e. $\frac{3}{4}$ of 5590 = 4192.5 revs/minute.

---

13. DISPERSION AND COLOUR

1. Show that $\dfrac{P_{1D}}{V_1} = \dfrac{P_{2D}}{V_2}$ and $P_D = P_{1D} - P_{2D}$ are the conditions

required for an achromatic prism.

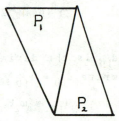

If the dispersion of the first
prism, figure 13.1, is equal
to that of the second prism,
then the combination will be
achromatic for the F and C
colours. That is, we require

Fig. 13.1

$$d_{1F} - d_{1C} = d_{2F} - d_{2C} \quad \dots\dots\dots(1)$$

where d is the prismatic deviation, the subscripts 1 and 2
refer to the prisms, and F and C refer to the Fraunhofer
blue and red lines using a hydrogen source.

For thin prisms $d = (n-1)a$. The n can be n_D, n_F, n_C, etcetera

and, where a subscript is not present, it usually implies the
D line (589.3nm, the middle of the sodium doublet) although
the helium D_3 or d line (587.56nm) is becoming the more common.

Now, for each prism and colour we have:

For P_1	For P_2
$d_{1D} = (n_{1D} - 1)a_1$	$d_{2D} = (n_{2D} - 1)a_2$
$d_{1C} = (n_{1C} - 1)a_1$	$d_{2C} = (n_{2C} - 1)a_2$
$d_{1F} = (n_{1F} - 1)a_1$	$d_{2F} = (n_{2F} - 1)a_2$

Hence, substituting in equation (1),

$$d_{1F} - d_{1C} = d_{2F} - d_{2C}$$
$$(n_{1F} - 1)a_1 - (n_{1C} - 1)a_1 = (n_{2F} - 1)a_2 - (n_{2C} - 1)a_2 ,$$

which reduces to $(n_{1F} - n_{1C})a_1 = (n_{2F} - n_{2C})a_2 \quad \dots\dots\dots(2)$

Substituting for $a_1 = d_{1D}/(n_{1D} - 1)$ and $a_2 = d_{2D}/(n_{2D} - 1)$

in equation (2), we have

$$\frac{n_{1F} - n_{1C}}{n_{1D} - 1} \cdot d_{1D} = \frac{n_{2F} - n_{2C}}{n_{2D} - 1} \cdot d_{2D}$$

or, $\omega_1 d_{1D} = \omega_2 d_{2D}$.

Now d_{1D} is the deviation produced by the thin prism so we can write $d_{1D} = P_{1D}$, and similarly for the second prism. Hence, $\omega_1 P_{1D} = \omega_2 P_{2D}$,   or $P_{1D}/V_1 = P_{2D}/V_2$, since $\omega = 1/V$.

The resultant prism will be $P_D = P_{1D} - P_{2D}$, giving the resultant in the same 'direction' as P_1 .

Dropping the D subscript, the conditions are

$$P = P_1 - P_2 , \qquad \text{and} \qquad \frac{P_1}{V_1} = \frac{P_2}{V_2}$$

---

2.   For a thin achromatic doublet show that the lenses must satisfy the conditions

$$F_D = F_{1D} + F_{2D} \qquad \text{and} \qquad \frac{F_{1D}}{V_1} = -\frac{F_{2D}}{V_2} .$$

Method 1

The resultant combination of two lenses (doublet) must bring C and F colours to the same focus. Hence, the power of the doublet for these colours must be the same; that is

$$F_F - F_C = 0 \quad \dots\dots\dots\dots\dots(1)$$

But, $F_F = F_{1F} + F_{2F}$ $\quad\dots\dots\dots\dots(2)$

and $F_C = F_{1C} + F_{2C}$ $\quad\dots\dots\dots\dots(3)$

Subtracting (3) from (2),

$$F_F - F_C = (F_{1F} - F_{1C}) + (F_{2F} - F_{2C}) = 0 \quad\dots\dots(4)$$

Now, in general, $F = (n-1)(R_1 - R_2) = (n-1)R$, having written $R = R_1 - R_2$. For the second lens in the doublet we shall write R' for the term $R_1 - R_2$. This saves carrying around a lot of curvature terms with subscripts. We can now write the following equations:

For the first lens	For the second lens

$$F_{1F} = (n_{1F} - 1)R \qquad\qquad F_{2F} = (n_{2F} - 1)R'$$

$$F_{1C} = (n_{1C} - 1)R \qquad\qquad F_{2C} = (n_{2C} - 1)R' \quad \Bigg] \quad \ldots\ldots\ldots(5)$$

$$F_{1D} = (n_{1D} - 1)R \qquad\qquad F_{2D} = (n_{2D} - 1)R'$$

Substituting from equation (5) into equation (4) rearranged,

$$F_{1F} - F_{1C} = -(F_{2F} - F_{2C})$$

gives $\qquad (n_{1F} - n_{1C})R = -(n_{2F} - n_{2C})R' \quad \ldots\ldots\ldots\ldots\ldots(6)$

But, from (5) again,

$$R = F_{1D}/(n_{1D} - 1)$$
$$\text{and} \quad R' = F_{2D}/(n_{2D} - 1) \quad \Bigg] \quad \ldots\ldots\ldots\ldots\ldots\ldots\ldots\ldots(7)$$

So, substituting from (7) into (6),

$$\frac{(n_{1F} - n_{1C})}{(n_{1D} - 1)} \cdot F_{1D} = -\frac{(n_{2F} - n_{2C})}{(n_{2D} - 1)} \cdot F_{2D} ,$$

which is $\qquad \dfrac{F_{1D}}{V_1} = -\dfrac{F_{2D}}{V_2} \qquad .$

Also, the combined power for the D line is the sum of the two thin lens powers for this line;

i.e. $\quad F_D = F_{1D} + F_{2D} .$

Dropping the D subscripts, the conditions are

$$\frac{F_1}{V_1} = -\frac{F_2}{V_2} \qquad \text{and} \qquad F = F_1 + F_2 .$$

Method 2

Consider the condition for the achromatic prism, $\frac{P_1}{V_1} = \frac{P_2}{V_2}$.

This equation has no sign convention applied, but if we agree that the base directions can be signified by + and − signs, then we can write

$$\frac{P_1}{V_1} = -\frac{P_2}{V_2} \ .$$

Now, Prentice's rule on a lens is $P = cF$, so if we consider the prismatic effect on an achromatic doublet at some point distance c above the principal axis, say, we can write

$$P_1 = cF_1 \quad \text{and} \quad P_2 = cF_2 \ ,$$

whence

$$\frac{cF_1}{V_1} = -\frac{cF_2}{V_2} \ ,$$

or

$$\frac{F_1}{V_1} = -\frac{F_2}{V_2} \ .$$

Again, $F_1 + F_2 = F$, the D subscripts being implied.

_____

3.   A convex lens is made using glass of index $n_d = 1.6$ and V-number 36.   It is found there is a difference of 0.25D between the powers for blue and red.   What is the power of the lens?

At a point c above the principal axis the prismatic effect is $d = cF$ prism dioptres of deviation.

Now, $\omega = \dfrac{d_F - d_C}{d_d} = \dfrac{cF_F - cF_C}{cF_d} = \dfrac{F_F - F_C}{F_d}$,

so $F_d = (F_F - F_C)/\omega = +0.25/(1/36) = +9D$.

_____

4. Assume a reduced standard emmetropic eye with radius of curvature 5.56mm. Express the axial (longitudinal) chromatic aberration in dioptres if the refractive index for red light is 1.332 and for blue light is 1.340.

Using the general expression for surface power,

$F = \dfrac{n' - n}{r}$, we have for blue (F) light $\quad F_F = \dfrac{1.340 - 1}{+0.00556} = +61.15D$,

and for red (C) light $F_C = \dfrac{1.332 - 1}{+0.00556} = +59.71D$.

Hence, the axial chromatic aberration is $F_F - F_C = 1.44D$

---

5. An achromatic thin prism deviates rays of D light by 5^0. It is made of crown and flint components with the following data:

	n_C	n_D	n_F
Crown	1.527	1.530	1.536
Flint	1.630	1.635	1.648

Find the apical angles of the component prisms.

The V-numbers are given by $\dfrac{n_D - 1}{n_F - n_C}$.

For crown we have $\quad V_1 = \dfrac{1.530 - 1}{1.536 - 1.527} = 58.89$

and for flint $\quad V_2 = \dfrac{1.635 - 1}{1.648 - 1.630} = 35.28$.

Hence, the crown and flint prism components, P_1 and P_2, are related by

$$\frac{P_1}{V_1} = \frac{P_2}{V_2},$$

or, $P_1 = \dfrac{V_1}{V_2} \cdot P_2 = \dfrac{58.89}{35.28} \cdot P_2$ ....(1)

The combined prism is $P = 100\tan5^0 = 8.75^\Delta$

and $\qquad\qquad P = 8.75 = P_1 - P_2$ ....(2)

whence $\qquad\qquad P_1 = 8.75 + P_2$ ....(3)

Since the L.H.S. of equations (1) and (3) are equal we can equate the R.H.S., thus

$$\frac{58.89}{35.28} \cdot P_2 = 8.75 + P_2 ,$$

and $P_2 = 8.75/(\frac{58.89}{35.28} - 1) = 13.07^\Delta .$

So, $P_1 = P + P_2 = 8.75 + 13.07 = 21.82^\Delta .$

Using $d = P = (n-1)a$ to relate apical angle a to prism power P, the apical angle for the crown component is

$$a_1 = P_1/(n_D - 1) = 21.82/(1.530 - 1) = 41.17^\Delta = 22.38^0 .$$

For the flint prism,

$$a_2 = P_2/(n_D - 1) = 13.07/(1.635 - 1) = 20.58^\Delta = 11.62^0 .$$

---

6. A thin prism with apical angle 8^0 has the following refractive indices:

$n_F = 1.514$, $n_D = 1.509$, $n_C = 1.506$.

Find the angular dispersion between the red and blue lines and the dispersive power of the glass.

Since $d = (n-1)a$, we have the angular separation of the C and F lines given by

$$d_F - d_C = (n_F - 1)a - (n_C - 1)a = (n_F - n_C)a = (1.514 - 1.506) \times 8^0$$
$$= 0.064^0 .$$

The dispersive power is

$$\omega = \frac{n_F - n_C}{n_D - 1} = \frac{1.514 - 1.506}{1.509 - 1} = 0.01572 .$$

---

7. The V-number of a +10D thin lens is 40. What is the axial chromatic aberration in dioptres?

At a point distance c cm above the principal axis the deviation of a ray is cF prism dioptres, from Prentice's rule. Hence, the deviations for the C, D, and F lines are

$$d_C = cF_C , \quad d_D = cF_D , \quad \text{and} \quad d_F = cF_F .$$

Now, $V = \dfrac{d_D}{d_F - d_C} = \dfrac{cF_D}{cF_F - cF_C} = \dfrac{F_D}{F_F - F_C} .$

But $F_F - F_C$ is the chromatic aberration, and we know V and F_D, so

$$F_F - F_C = \frac{F_D}{V} = \frac{+10}{40} = +0.25D$$

---

8. A cemented thin achromatic lens of power +4D is constructed from two glasses of indices 1.50 and 1.625 and dispersive powers 1/60 and 1/36, respectively. The outer surface of the flint lens is plane. Calculate the radii of the other surfaces.

Let F_1 be the crown lens and F_2 the flint lens. Then

$$F = 4 = F_1 + F_2 \quad \ldots\ldots\ldots\ldots\ldots\ldots\ldots(1)$$

and

$$F_1 = -\frac{V_1}{V_2} \cdot F_2 \quad \ldots\ldots\ldots\ldots\ldots\ldots\ldots\ldots(2)$$

Substituting for F_1 from equation (2) into equation (1), we have

$$F_2 - \frac{V_1}{V_2} \cdot F_2 = 4 \ ,$$

or

$$F_2 = 4/(1 - \frac{60}{36}) = -6D,$$

having used $V_1 = 1/\omega_1 = \frac{1}{1/60} = 60$ and $V_2 = 1/\omega_2 = \frac{1}{1/36} = 36$.

So, putting $F_2 = -6$ in equation (2),

$$F_1 = -\frac{V_1}{V_2} \cdot F_2 = -\frac{60}{36}(-6) = +10D \ .$$

Since the flint lens is -6D and one surface is plane, all the power is on the contact surface. Hence, the radius of this surface is

$$r = \frac{n' - n}{F} = \frac{1.625 - 1}{-6} = -0.1042m = -10.42cm,$$

imagining the flint lens in air.

The power of this surface for the crown lens is

$$F = \frac{n' - n}{r} = \frac{1 - 1.50}{-0.1042} = +4.80D \ _o$$

Since $F_1 = +10D$, the first surface is $10 - 4.8 = +5.2D$, and the radius of this surface is

$$r = \frac{n' - n}{F} = \frac{1.50 - 1}{5.2} = +0.0962m = +9.62cm.$$

---

9. A direct vision spectroscope contains two flint prisms and three crown prisms, the latter being 10^0 refracting angles. Given the data below, calculate the refracting angle of the identical flint prisms and the dispersion produced by the instrument.

	n_C	n_D	n_F
Crown glass	1.515	1.518	1.524
Flint glass	1.618	1.623	1.635

The D line must suffer no deviation on leaving the spectoscope We must find the total deviation produced by the three crown prisms. Each prism produces $d_D = (n_D - 1)a$ deviation. Putting $n_D = 1.518$ and $a = 10^0$, we have

$$d_D = (1.518 - 1) \times 10^0 = 5.18^0.$$

Thus, the total deviation is 15.54^0. Opposed to this are two flint prisms, each of which must produce $d_D = 15.54^0/2 = 7.77^0$

Hence, the refracting angle of each flint prism is

$$a = d_D/(n_D - 1) = 7.77/(1.623 - 1) = 12.47^0.$$

Of course, the crown and flint prisms are alternate in the instrument.

To find the dispersion we must first find the deviations for red and blue for each type of prism.

For a crown prism

$d_C = (n_C - 1)a = (1.515 - 1) \times 10^0 = 5.15^0$, and

$d_F = (n_F - 1)a = (1.524 - 1) \times 10^0 = 5.24^0.$

For a flint prism

$d_C = (n_C - 1)a = (1.618 - 1) \times 12.47^0 = 7.71^0$, and

$d_F = (n_F - 1)a = (1.635 - 1) \times 12.47^0 = 7.92^0.$

Now, the total deviation of the red light is
$$3 \times d_C \text{(for crown)} - 2 \times d_C \text{(for flint)}$$
$$= (3 \times 5.15^0) - (2 \times 7.71^0)$$
$$= 0.03^0.$$

The total deviation for blue light is
$$(3 \times 5.24^0) - (2 \times 7.92^0)$$
$$= -0.12^0.$$

The minus sign here indicates the resultant deviation of the blue light is in the opposite direction to that of the red light. Hence, the angular separation is

$$0.03^0 - (-0.12^0) = 0.15^0.$$

To obtain a feel for this angular size remember the moon subtends about 0.5^0 at the Earth.

---

10. An equiconvex lens and an equiconcave lens are placed in contact to form an achromatic combination. Calculate the curvature of the surfaces of the concave lens and the power of the combination given

	n_C	n_D	n_F
Concave lens	1.610	1.639	1.667
Convex lens	1.480	1.490	1.500

and the curvature of each surface of the convex lens is numerically 10D.

We can calculate the chromatic aberration of the convex lens. Thus,

$$F_F - F_C = (n_F - 1)(R_1 - R_2) - (n_C - 1)(R_1 - R_2)$$

$$= (n_F - n_C)(R_1 - R_2)$$

$$= (1.500 - 1.480)(10 - (-10))$$

$$= +0.40D$$

The concave lens must have a chromatic aberration equal and opposite; i.e. -0.40D. We can therefore find the curvature of the surfaces from the relationship in the last equation.

$$F_F - F_C = (n_F - n_C)(R_1 - R_2) .$$

Thus, $R_1 - R_2 = (F_F - F_C)/(n_F - n_C) = -0.40/(1.667 - 1.610) = -7.02D$

But $R_1 = -R_2$, for an equiconcave lens,
so $2R_1 = -7.02D$, and $R_1 = -3.51D$ and $R_2 = +3.51D$.

To find the power of the combination we must find F_D for each lens.

F_D for the convex lens:

$$F_D = (n_D - 1)(R_1 - R_2) = (1.490 - 1)(10 - (-10)) = +9.80D$$

F_D for the concave lens:

$$F_D = (\eta_D - 1)(R_1 - R_2) = (1.639 - 1)(-3.51 - (+3.51)) = -4.49D.$$

The power of the combination is the sum of these two powers; i.e. +9.80 + (-4.49) = +5.31D.

---

11. A prism ABCD (A=90°, B=75°, C=135°, D=60°) of a constant deviation spectrometer is made of glass of refractive index $\eta_D = 1.646$. F is a point on the face AB. A ray of light EF from the collimator is refracted at AB, totally internally reflected at BC, refracted at AD, and emerges perpendicular to its original direction EF. Trace the ray through the prism.

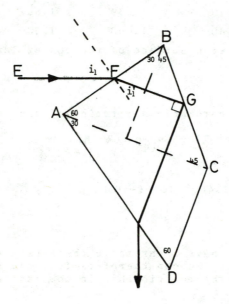

Fig. 13.2 The Pellin-Broca constant deviation prism.

If hydrogen light is transmitted by the collimator, find the dispersive power of the glass given that when hydrogen F (blue-green) is focused in the telescope the angle AFE is 34°4' , and when hydrogen C (red) is focused the prism has to be rotated through an angle 0°56'.

The dispersive power of the glass is given by $\omega = (n_F - n_C)/(n_D - 1)$.
The D refers to the Fraunhofer mean sodium wavelength, 589.3nm.
Clearly, we must find n_F and n_C. We already know $n_D = 1.646$.
The ray FG is parallel to the line AC in this particular prism,
so $\qquad i_1' = 90^0 - B\hat{F}G = 90^0 - 60^0 = 30^0$,

and $\qquad i_1 = 90^0 - A\hat{F}E = 90^0 - 34^0 \; 4' = 55^0 \; 56'$.

Hence, $\qquad n_F = \dfrac{\sin i_1}{\sin i_1'} = \dfrac{\sin 55^0 \; 56'}{\sin 30^0} = 1.657$.

When red light enters the face AB i_1' must again be 30^0 and
since the red light will be deviated less than the blue i_1
must reduce by the $0^0 \; 56'$ through which the prism is rotated.

So, $\qquad n_C = \dfrac{\sin 55^0}{\sin 30^0} = 1.638$,

and $\qquad \omega = \dfrac{n_F - n_C}{n_D - 1} = \dfrac{1.657 - 1.638}{1.646 - 1} = 0.0294$.

14. INTERFERENCE

Wavefront-splitting interferometer questions

1. What are the conditions necessary to produce sustained
    interference between two waves? In a Young's double slit
    arrangement the screen is 1.0m away from the double slits
    which are 1.8mm apart. If light of 546nm is used find the
    separation of the fringes on the screen.

To produce a sustained interference pattern the two waves
must overlap and be coherent. The waves will be coherent
if the phase difference is constant. Suppose two waves
propagate along the x-axis with equal wavelength (and therefore
frequency); let the waves be

$$E_1 = E_{01} \cos(kx - \omega t + \varepsilon_1)$$

and $\quad E_2 = E_{02} \cos(kx - \omega t + \varepsilon_2)$, where E_{01} and E_{02} are the
amplitudes, $k = 2\pi/\lambda$, and $\omega = 2\pi\nu$. λ is the wavelength and
is the frequency.

Then the phases are $\phi_1 = kx - \omega t + \varepsilon_1$

$$\text{and } \phi_2 = kx - \omega t + \varepsilon_2 \, ,$$

and the phase difference is $\phi_1 - \phi_2 = \varepsilon_1 - \varepsilon_2$, and this must be
constant.

In practice waves are not infinitely long as are the two waves
above. Hence, we must ensure that their optical path lengths
do not differ by more than the wave-train length, or coherence
length, otherwise they will not arrive at some point P, on a
screen say, during the time interval in which they can interfere.

One further point, for the fringes to appear the amplitude
vectors must be parallel. For example, two plane-polarised
waves which have their vectors mutually perpendicular cannot
produce maxima and minima; Hence, no interference fringe
pattern is discernible.

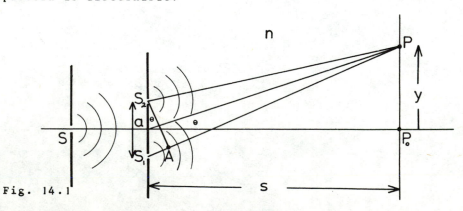

Fig. 14.1

Figure 14.1 shows Young's experiment. S is a slit perpendicular to the page and S_1 and S_2 are two slits separated by a distance a. P_0 is a point on the screen such that SP_0 may be regarded as the axis of the system.

Waves arising from S arrive at S_1 and S_2 with a constant phase difference which will be zero if $SS_1 = SS_2$. The secondary waves arising from S_1 and S_2 are therefore coherent. Although the waves arising from S have finite wave-train lengths, or coherence lengths, $\Delta\ell$, if the optical path difference, $n.S_1A$ is less than $\Delta\ell$ then interference will occur at P. For a maximum at P the optical path difference(O.P.D.) must be an exact number of (vacuum) wavelengths;

i.e. $n.S_1A = m\lambda$, where $m = 0, \pm1, \pm2,\ldots.$

But, from the figure, $S_1A = a\sin\theta = a\theta$ if θ is small.

Also, $\sin\theta \simeq \tan\theta \simeq \theta = y/s$ if θ is small, which it is if $s \gg y$.

Hence, substituting for θ, we have the mth bright fringe at a distance y_m from P_0 , given by

$$n.S_1A = na\frac{y_m}{s} = m\lambda$$

or $y_m = \dfrac{sm\lambda}{na}$.

Now, the (m+1)th and the mth bright fringes are separated by

$$\Delta y = y_{m+1} - y_m = \frac{s(m+1)\lambda}{na} - \frac{sm\lambda}{na} = \frac{s\lambda}{na} .$$

If the apparatus is in air n=1. Then in the case where $\lambda = 546 \times 10^{-9}$m, $a = 1.8 \times 10^{-3}$m, and s = 1.0m, the fringe separation (sometimes called the fringe width) is

$$\Delta y = \frac{s\lambda}{na} = \frac{1.0 \times 546 \times 10^{-9}}{1 \times 1.8 \times 10^{-3}} = 303.3 \times 10^{-6}\text{m}$$

$$= 0.303\text{mm}$$

---

2. In question 1 a glass chamber filled with air is placed before the upper slit, S_2. The air is then replaced with a gas when the fringe pattern is seen to be displaced by 14 bright bands. If the chamber is 16mm long find the refractive index of the gas, n_g, given the refractive index of air is 1.000275. In which direction is the fringe pattern moved?

The geometric path length of the chamber is 16×10^{-3} m and when containing air or gas the optical path length is $n_a \times 16 \times 10^{-3}$m and $n_g \times 16 \times 10^{-3}$m , respectively. Replacing the air by gas changes the optical path length and the optical path difference is

$$O.P.D. = (n_g - n_a) \times 16 \times 10^{-3} m.$$

This is equal to 14 (vacuum) wavelengths,
so
$$14 \times 546 \times 10^{-9} = (n_g - 1.000275) \times 16 \times 10^{-3},$$

whence $\quad n_g = \dfrac{14 \times 546 \times 10^{-9}}{16 \times 10^{-3}} + 1.000275 = 1.000753$.

Since the 'upper' path length is increased (the O.P.D.> 0) the fringe pattern moves downwards.

---

3. In Young's experiment with light of wavelength 587.5nm it was found that the separation of the first and eleventh dark fringes was 4.44mm at a distance of 2m from the slits. (i) Find the separation of the slits. (ii) What would be the separation of the first and eleventh bright fringes if the apparatus were immersed in water of refractive index 1.333? (iii) How would you find the central bright fringe?

(i) A dark fringe will occur when the O.P.D. is a whole number of wavelengths plus one half-wavelength: the two waves, having equal amplitudes, now superimpose with crest on trough and destructively interfere. Hence, the mth dark fringe occurs when
$$O.P.D. = \frac{na y_m}{s} = (m+\tfrac{1}{2})\lambda \ , \quad m = 0, \pm 1, \pm 2, \ldots$$

The distance between two adjacent dark fringes is
$$\Delta y = y_{m+1} - y_m = \frac{s(m+1+\tfrac{1}{2})\lambda}{na} - \frac{s(m+\tfrac{1}{2})\lambda}{na} = \frac{s\lambda}{na},$$
as for bright fringes!

Between the first and eleventh fringes there are ten fringe widths , so
$$10.\Delta y = 10\frac{s\lambda}{na} \ .$$

When n=1 $\quad 10.\Delta y = 4.44 \times 10^{-3} m$, whence the separation of the slits is
$$a = \frac{10s\lambda}{n \times 10\Delta y} = \frac{10 \times 2 \times 587.5 \times 10^{-9}}{1 \times 4.44 \times 10^{-3}} = 2650 \times 10^{-6} m$$

$$= 2.65mm.$$

(ii) The first and eleventh bright fringes would have 10 fringe width separations in water. Thus, we have
$$10.\Delta y = 10\frac{s\lambda}{na} = \frac{10 \times 2 \times 587.5 \times 10^{-9}}{1.333 \times 2.65 \times 10^{-3}} = 3330 \times 10^{-6} m$$

$$= 3.33mm.$$

This is the same as 4.44/1.333; that is, the fringe width in air divided by the refractive index of water.

(iii) To find the position of the central bright fringe illuminate the slits with white light when the central fringe will be the only white fringe.

---

4. Explain how interference fringes are obtained with a Fresnel biprism. If the apical angles are 0.01 radian, the prism is 10cm from the slit and the screen is 90cm from the prism, calculate the fringe width for light of 600nm wavelength. The prism glass has refractive index $n_p = 1.5$.

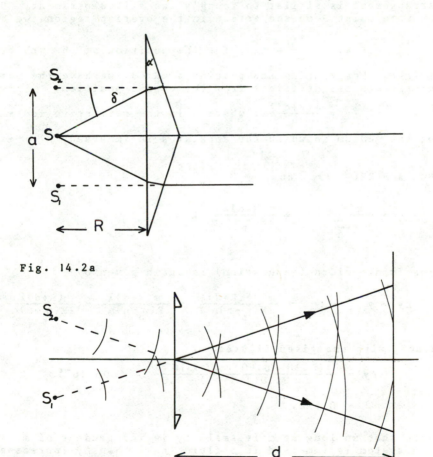

Fig. 14.2a

Fig. 14.2b

An illuminated slit, S, parallel to the common base of the biprism, acts as a source. From figure 14.2a it can be seen to produce two secondary sources S_1 and S_2. The angular deviation produced by each half of the (thin) biprism is $\delta = (\frac{n_p}{n} - 1)\alpha$, where α is the apical angle of each half and n is the refractive index in which the prism is immersed. From the figure the distance between S_1 and S_2 is given by

$$a = 2R\tan\delta \simeq 2R\delta, \text{ since } \tan\delta \simeq \delta, \text{ with } \delta \text{ in radians.}$$

Thus, using $\delta = (\frac{n_p}{n} - 1)\alpha$, $\quad a = 2R(\frac{n_p}{n} - 1)\alpha$.

Where the waves from S_1 and S_2 overlap to the right of the biprism, figure 14.2b, interference fringes occur. The arrangement is similar to Young's two slit experiment. Hence, at some point P on the screen in the overlap region, we have

$$O.P.D. = \frac{nay_m}{s} = m\lambda \text{ , for the position of the mth bright}$$

fringe. Rearranging and putting $s = R + d$, we have the position of the mth bright fringe from the cenral position,

$$y_m = \frac{m(R+d)\lambda}{na} \text{ , where n is the refractive index}$$

of the medium in which the apparatus is immersed.

But $a = 2R(\frac{n_p}{n} - 1)\alpha$, so

$$y_m = \frac{m(R+d)\lambda}{2nR(\frac{n_p}{n} - 1)\alpha} = \frac{m(R+d)\lambda}{2R(n_p - n)\alpha} \text{ .}$$

The fringe width (separation) is again given by

$$\Delta y = y_{m+1} - y_m = \frac{(m+1)(R+d)\lambda}{2R(n_p - n)\alpha} - \frac{m(R+d)\lambda}{2R(n_p - n)\alpha} = \frac{(R+d)\lambda}{2R(n_p - n)\alpha} \text{ .}$$

Hence, with the given figures,

$$\Delta y = \frac{(0.10 + 0.90)\times600 \times 10^{-9}}{2 \times 0.10(1.5 - 1) \times 0.01} = 600 \times 10^{-6}\text{m}$$

$$= 0.6\text{mm.}$$

Note that so long as α is small Δy is independent of m. If the system is immersed in a fluid ($n > 1$) then Δy increases.

5. A collimated laser beam impinges directly on a Fresnel biprism. Show that the fringe width is independent of the position of the screen.

A collimated laser beam is effectively parallel light emerging from a point source at a great distance. Hence $R \gg d$ and $R+d \simeq R$, the terms being as in the last question.

So, $\Delta y = \dfrac{(R+d)\lambda}{2R(n_p - 1)\alpha}$, for the system in air with n=1,

$\qquad = \dfrac{\lambda}{2(n_p - 1)\alpha}$, since $R+d \simeq R$.

Clearly, this is independent of d. Suppose the laser is a helium-neon laser ($\lambda = 632.8$nm), $\alpha = 0.5^{\circ} = 8.727 \times 10^{-3}$ radian, then the fringe separation is

$\qquad \Delta y = \dfrac{632.8 \times 10^{-9}}{2(1.5 - 1) \times 8.727 \times 10^{-3}} = 72.51 \times 10^{-6}$m

$\qquad\qquad = 0.07251$mm.

AMPLITUDE SPLITTING - THIN FILM PROBLEMS

6. Explain how interference fringes are produced by thin films.

Firstly, consider the phase and amplitude of reflected light. For NORMAL incidence the fractional amplitude of the reflected wave is given by Fresnel's equation

$\qquad\qquad -(\dfrac{n' - n}{n' + n})$.

That is, if the amplitude of the normally incident wave is A_i then the amplitude of the reflected wave is

$\qquad\qquad A_r = -(\dfrac{n' - n}{n' + n})A_i$.

Note the R.H.S. is negative if $n' > n$ and positive if $n' < n$. The negative sign indicates a change of phase of π radians, equivalent to a path difference of $\lambda/2$, when light is reflected from a medium of higher index.

Secondly, the fraction of the intensity reflected is given by the square of the fraction of the amplitude reflected. The fractional intensity reflected, called the reflectance, is given by

$\qquad\qquad R = (-(\dfrac{n' - n}{n' + n}))^2 = (\dfrac{n' - n}{n' + n})^2$.

In order to easily distinguish fringes the two interfering
waves should have approximately equal amplitudes, as they
were in the wavefront splitting problems. This means the
reflectances for the two surfaces of the film should be
approximately equal. Multiple reflections, which in any
case have much lower amplitudes, are ignored in treatments
of single films.

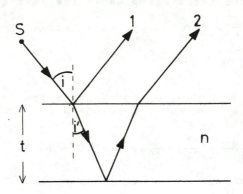

Fig. 14.3

Consider a ray arising from a point source, S, in figure
14.3. The amplitude of the reflected waves, 1 and 2, will
be nearly equal and can be made to overlap and interfere by
by being brought to a focus on a screen. Often the eye's
lens and retina suffice for this purpose. It can be shown
that the O.P.D. for rays 1 and 2 is given by $2nt\cos i'$, when
neither ray suffers or both rays undergo a phase change of π,
or by $(2nt\cos i' - \lambda/2)$ when one ray undergoes a π phase change.

Since the O.P.D. depends on t and i', and not on the position
of the point source S, an extended source can be used. In
that case other rays contribute to the fringes.

Two particular fringe pattern types are given special names:

(i) fringes of equal inclination, when t is constant (a flat,
parallel sided film), are called HAIDINGER fringes. The
fringe pattern is circular.

(ii) fringes of equal thickness, when i is effectively constant
but t varies, are called FIZEAU fringes. Patterns of fringes
of constant thickness map out regions of space between surfaces
where the thickness is equal in much the same way as contour
lines signify regions of equal height on a map.

7. State the conditions for dark and bright fringes in a thin
    film of refractive index n. A thin transparent film of
    refractive index 1.432 is to generate a minimum in reflected
    light, $\lambda = 500nm$, under normal incidence. Find the minimum
    thickness of the film to achieve this.

Assuming the media on each side of the film are identical
then one reflected ray will suffer π radians phase change.
Hence, the O.P.D. $= 2ntcos\ i' - \lambda/2$.

For bright fringes the reflected rays, 1 and 2 in figure 14.3,
must be in phase. This is achieved when the O.P.D. is equal
to a whole number of wavelengths.

Stated mathematically this is

$$O.P.D. = 2ntcos\ i' - \lambda/2 = m\lambda$$

$$or,\ \ 2ntcos\ i' = (m+\tfrac{1}{2})\lambda \ \ \ \ , \ m = 0,1,2,....$$

Dark fringes occur when the O.P.D. differs from the condition
for bright fringes by $\lambda/2$.
That is
$$2ntcos\ i' = m\lambda \ .$$

For a finite minimum thickness of film put m=1 in the
condition for a dark fringe above, which gives

$$t = \frac{\lambda}{2n} \ \ \ , \ since\ cos\ i' = 1 \ \ (i=i'=0).$$

For $\lambda = 500nm$ and n = 1.432 we have $t = \dfrac{500}{2 \times 1.432} = 174.6nm.$

---

8. Show that a film can be made 'antireflecting' if $n_f = \sqrt{n_1 n_3}$
    and $t = \lambda/4n_f$ with the data as in figure 14.4.
    Assume $n_1 < n_2 < n_3$.

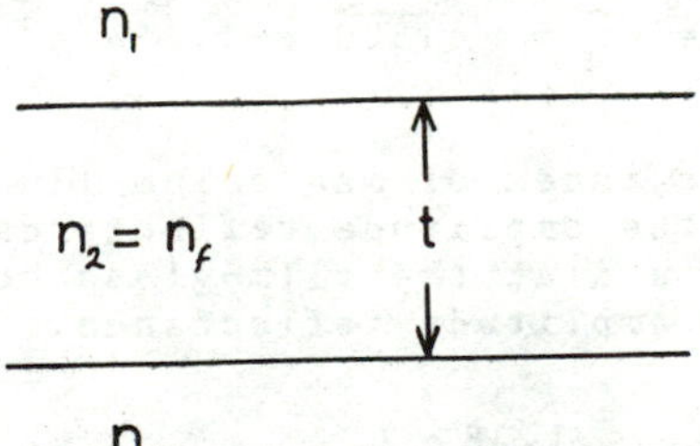

Fig. 14.4

The amplitudes of the reflected waves from the top and bottom
surfaces must be equal. This will be so when

$$-\left(\frac{n_f - n_1}{n_f + n_1}\right) = -\left(\frac{n_3 - n_f}{n_3 + n_f}\right) .$$

Removing the brackets and rearranging gives

$$n_f n_3 + n_f^2 - n_1 n_3 - n_1 n_f = n_1 n_3 - n_f^2 + n_1 n_3 - n_1 n_f \ ,$$

which reduces to $\quad n_f^2 = n_1 n_3$

or $\qquad n_f = \sqrt{n_1 n_3} \ .$

Now, the O.P.D. $= 2 n_f t \cos i'$ and for a minimum we have

$$2 n_f t \cos i' = (m + \tfrac{1}{2}) \lambda$$

(since both rays suffered π phase change the O.P.D. must be an odd number of half-wavelengths for the reflected waves to be antiphase).

For normal incidence $\cos i' = 1$ and for minimum thickness $m = 0$, so

$$t = \frac{\tfrac{1}{2}\lambda}{2 n_f} = \frac{\lambda}{4 n_f} \ .$$

Suppose $n_1 = 1$ (air) and $n_3 = n_s$, where s stands for substrate, then the conditions for an antireflection coating on a lens surface are

$$t = \lambda / 4 n_f \quad \text{and} \quad n_f = \sqrt{n_s} \ .$$

---

9. Magnesium fluoride, index 1.38, is used as an antireflection film on the surfaces of a crown glass lens, index 1.523. Show that this film cannot reduce the reflected light to zero and find what percentage is reflected.

For the amplitude of the two reflected waves to be equal we must have

$$n_f = \sqrt{n_s} = \sqrt{1.523} = 1.234 \ .$$

The refractive index of magnesium fluoride is too high! If we calculate the amplitude reflectances at the air-film boundary, r_1 , and at the film-glass boundary, r_2 , we can find the resultant amplitude reflectance.

$$r_1 = -\left(\frac{n_f - n_1}{n_f + n_1} \right) = -\left(\frac{1.38 - 1}{1.38 + 1} \right) = -0.160 \quad \text{and}$$

$$r_2 = -\left(\frac{n_s - n_f}{n_s + n_f} \right) = -\left(\frac{1.523 - 1.38}{1.523 + 1.38} \right) = -0.049 \ .$$

For a quarter-wave film, that is $t = \lambda/4n_f$, the reflected
waves are in antiphase (crest superimposing on trough) so
that the resultant amplitude reflectance is

$$r_1 - r_2 = -0.160 - (-0.049) = -0.111 \quad .$$

The intensity reflectance is

$$(r_1 - r_2)^2 = (-0.111)^2 = 0.012 \quad \text{or} \quad 1.2\% \quad .$$

---

10. Suppose a thin wedge-shaped film is formed between two flat
glass plates. For nearly normal incidence find expressions
for the fringe separation, the position of the mth maximum
measured from the apex of the wedge, and the film thickness
at the mth bright fringe. Describe the pattern formed.
Two thin, flat glass plates are separated by a hair, thus
forming a wedge-shaped air-film. Find the diameter of the
hair if it lies in the position of the 345th bright fringe
when the wedge is illuminated by sodium light ($\lambda = 589.3$nm)
under normal incidence.

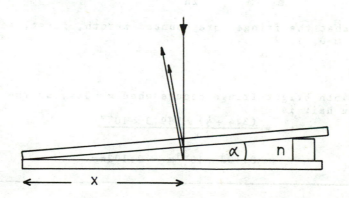

Fig. 14.5

Since one reflection is at a glass-film boundary and the
other at a film-glass boundary there will be a 'hidden' phase
change equivalent to $\lambda/2$. The condition for maxima is

$$2nt \cos i' - \frac{\lambda}{2} = m\lambda , \quad m = 0, 1, 2, \ldots\ldots$$

or
$$2nt = (m+\tfrac{1}{2})\lambda , \quad \text{since} \cos i' = 1 \quad .$$

But, from the triangular wedge we have for the thickness of
the film at the mth maximum

$$t_m = x_m \tan \alpha \simeq x_m \alpha , \quad \text{where } \alpha \text{ is the wedge-angle.}$$

Substituting for t in the condition for maxima,

$$2nx_m\alpha = (m+\tfrac{1}{2})\lambda \ .$$

Rearranging this gives the position of the mth maximum

$$x_m = \frac{(m+\tfrac{1}{2})\lambda}{2n\alpha} \ .$$

The fringe width is given by

$$\Delta x = x_{m+1} - x_m = \frac{(m+1+\tfrac{1}{2})\lambda}{2n\alpha} - \frac{(m+\tfrac{1}{2})\lambda}{2n\alpha} = \frac{\lambda}{2n\alpha} \ .$$

The fringes formed are fringes of equal thickness. That is, the mth bright fringe forms where the wedge is t_m thick and this will be a line parallel to the apex of the wedge. Hence, the pattern is a series of straight alternate dark and bright bands. At the apex the glass plates are in contact and no reflection takes place, so it is dark. The mth bright fringe occurs when t is given by

$$t_m = x_m\alpha = \frac{(m+\tfrac{1}{2})\lambda}{2n} \ , \text{ using the equation for } x_m \ .$$

Note that the fringes are counted zeroth, first, second, ..., since m=0, 1, 2,.......

The 345th bright fringe occurs when m = 344, so the diameter of the hair is

$$t_{344} = \frac{(344 + \tfrac{1}{2}) \times 589.3 \times 10^{-9}}{2 \times 1}$$

$$= 0.102 \times 10^{-3}m = 0.102mm.$$

---

11. Newton's rings are formed by reflection at the film formed between the convex surface of a lens and the flat, glass plate upon which it rests. When sodium light, wavelength 589.3nm, is used the diameter of the second dark ring is 0.238cm and that of the twenty-second is 0.788cm. What is the radius of curvature of the convex surface?

For nearly normal incidence the geometric path difference between the two reflected rays is 2s, very nearly. If the refractive index of the film is n, minima will occur when the O.P.D. is a whole number of wavelengths, bearing in mind there is a hidden half-wavelength shift at the second reflection.

Hence, for dark fringes $2ns = m\lambda$.

Bright fringes therefore occur when $2ns = (m+\frac{1}{2})\lambda$.

Using the approximate sag relationship $s = y^2/2r$ where the symbols have their usual meaning, we substitute for s in the equation for dark fringes.

Thus
$$2n \cdot y^2/2r = m\lambda \; , \; \text{which gives}$$

$$r = ny^2/m\lambda \; , \quad m = 0, \; 1, \; 2, \dots$$

Rearranging this, $m\lambda r = ny^2$, and if the radius of the dark ring is y_1 when the order is m_1, and similarly with y_2 and m_2, we have
$$m_1\lambda r = ny_1^2 \quad \dots\dots\dots\dots(1)$$

$$m_2\lambda r = ny_2^2 \quad \dots\dots\dots\dots(2)$$

Subtracting (2) from (1),
$$(m_1 - m_2)\lambda r = n(y_1^2 - y_2^2)$$

or
$$r = \frac{n(y_1^2 - y_2^2)}{(m_1 - m_2)\lambda}$$

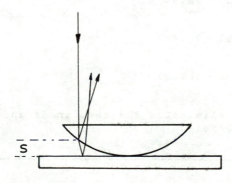

Fig. 14.6

So the radius of the convex surface is
$$r = \frac{1 \times (0.394^2 - 0.119^2)}{(22 - 2) \times 589.3 \times 10^{-7}}$$

$$= 120 \text{cm},$$

where $y_2 = 0.788/2$cm, $y_1 = 0.238/2$cm, and $\lambda = 589.3 \times 10^{-7}$cm.

Note in passing that since $y^2 = m\lambda r/n$, increasing n by placing water, say, between the lens and the plate instead of air, will cause y to decrease for each m. That is, the pattern shrinks towards the centre.

---

12 Suppose some dust separates the lens and the plate by an unknown distance Δs. Show that the previous method still works.

The O.P.D. is now $2n(s+\Delta s)$ since s is effectively increased by Δs. Hence, the mth dark fringe occurs when

$$2n(s+\Delta s) = m\lambda .$$

Putting $s = y^2/2r$ gives

$$\frac{2ny^2}{2r} + 2n.\Delta s = m\lambda$$

which rearranges and simplifies to

$$y^2 = \frac{r}{n}(m\lambda - 2n.\Delta s) .$$

So
$$y_1^2 = \frac{r}{n}(m_1\lambda - 2n.\Delta s)$$

and
$$y_2^2 = \frac{r}{n}(m_2\lambda - 2n.\Delta s) .$$

Subtracting these eliminates the term in Δs and the expression for r follows as before.

---

13. Describe the Michelson interferometer and its underlying principles. The device is set up to show circular fringes. Suppose the mirror, M_1, the moveable mirror, is initially a distance d further from the beam splitter than is mirror M_2. As d is reduced fringes sweep towards the centre of the field of view. If 800 bright fringes pass by when d is reduced by 1.20×10^{-4}m, find the wavelength of the monochromatic light used.

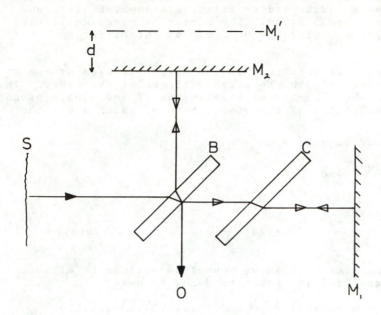

Fig. 14.7

A beam from an extended source is amplitude split at the
beam-splitter B. One wave is reflected from the mirror M_2
and split again at B, half going to the observer at 0 and
half going back to the source S. Of the initial beam striking
B the other half traverses the compensator plate, C, is
reflected at M_1 and half of this reaches the observer after
reflection at B.

The observer sees the mirror M_2 through the beam-splitter and
the image of M_1, that is M_1', by reflection in B. The observer
effectively sees two mutually coherent beams, one coming from
M_2 and one from M_1'. If M_1 and M_2 are perpendicular then M_1' and
M_2 are parallel and the two beams are to all intents reflected
from a parallel sided plate. In monochromatic light this
results in circular Haidinger fringes of equal inclination.

If M_1' and M_2 are slightly inclined the effect is that of a wedge
and straight Fizeau fringes of equal thickness are formed.

The O.P.D. of the two beams is 2nd, since d is traversed
twice. This is 2d since n = 1 (air). Each fringe passing by
corresponds to a movement of M_1 by $\lambda/2$. So

$$800 \times \frac{\lambda}{2} = 2 \times 1.20 \times 10^{-4}$$

or $\lambda = \dfrac{2 \times 2 \times 1.20 \times 10^{-4}}{800} = 6.00 \times 10^{-7} \text{m} = 600 \text{nm}.$

14. A Michelson interferometer is set to give maximum visibility of fringes for a sodium source emitting a doublet with wavelengths 589.0nm and 589.6nm. The mirror is moved until the fringes disappear. Find the movement of the mirror.

The visibility will be high when the bright fringes of one wavelength coincide with the bright fringes of the other. This occurs when a whole number of wavelengths of one equals a whole number of wavelengths of the other, both of which equal the O.P.D.

That is   O.P.D. $= 2d = m_1\lambda_1 = m_2\lambda_2$.

Hence, $m_1 = \dfrac{2d}{\lambda_1}$ and $m_2 = \dfrac{2d}{\lambda_2}$.

Subtracting gives

$$m_1 - m_2 = \frac{2d}{\lambda_1} - \frac{2d}{\lambda_2} = 2d(\frac{\lambda_2 - \lambda_1}{\lambda_1\lambda_2}) \quad \dots\dots\dots\dots(1)$$

Now, $m_1 - m_2$ increases by $\frac{1}{2}$ as we move from maximum to minimum visibility, and if d increases to d+Δd, we have

$$m_1 - m_2 + \tfrac{1}{2} = 2(d+\Delta d)(\frac{\lambda_2 - \lambda_1}{\lambda_1\lambda_2}) \quad \dots\dots\dots\dots\dots(2)$$

Subtracting (1) from (2) gives

$$\tfrac{1}{2} = 2.\Delta d(\frac{\lambda_2 - \lambda_1}{\lambda_1\lambda_2}),$$

or $\Delta d = \dfrac{\lambda_1\lambda_2}{4(\lambda_2 - \lambda_1)}$.

With the given figures the mirror moves

$$\Delta d = \frac{589.6 \times 589.0}{4(589.6 - 589.0)} = 1.447 \times 10^5 \text{nm}$$

$$= 0.1447\text{mm}.$$

15. A thin sheet of CR39 material, index 1.498, is inserted normally in one arm of a Michelson interferometer. Using a source emitting light of wavelength 600nm, 105 fringes are displaced. What is the sheet's thickness?

When inserting the sheet the optical path length changes from $1 \times t$ to 1.498t, where 1 is the refractive index of air and 1.498 is the refractive index of the sheet.
Hence, the O.P.D. $= 2(1.498 - 1)t$ since the light traverses the sheet twice. This must equal 105 wavelengths.
Thus

$$105\lambda = 2(1.498 - 1)t,$$

so, $t = \dfrac{105 \times 600 \times 10^{-9}}{2(1.498 - 1)}$ m $= 6.325 \times 10^{-2}$mm.

15. DIFFRACTION

1. Plane waves of 550nm wavelength are incident normally on
   a narrow slit of width b=0.25mm. Calculate the distance
   between the first minima on either side of the axis when the
   Fraunhofer diffraction pattern is imaged by a lens of focal
   length 60cm. The lens should be 'large'. Why is this?

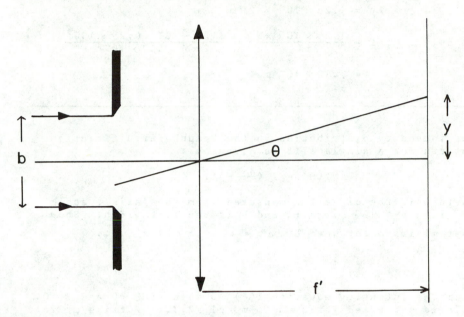

Fig. 15.1

The distance from the central maximum to the first minimum
on one side is given by $y = f'\tan\theta \simeq f'\sin\theta$, for small θ.
Minima occur for $b\sin\theta = m\lambda$, $m = \pm1, \pm2, \ldots$ Substituting for
$\sin\theta$ from the second equation into the first gives

$$y = f'm\lambda/b = 60 \times 10^{-2} \times 1 \times 550 \times 10^{-9}/0.25 \times 10^{-3}$$

$$= 1.32 \times 10^{-3}\text{m} = 1.32\text{mm}.$$

But, the distance between the two first order minima is
$2y = 2.64$mm.

The lens must be large compared to the slit in order not
to introduce its own diffraction effects.

2. Plane waves (λ=550nm) fall normally on a slit 0.25mm wide. The separation of the fourth-order minima of the Fraunhofer diffraction pattern in the focal plane of a converging lens is 1.25mm. Calculate the focal length of the lens.

From the last question, with $m = 4$, we have the distance between the two fourth-order minima is

$$2y = 1.25 \times 10^{-3} = 2f'm\lambda/b \ .$$

So,
$$f' = \frac{1.25 \times 10^{-3} \times b}{2m\lambda} = \frac{1.25 \times 10^{-3} \times 0.25 \times 10^{-3}}{2 \times 4 \times 550 \times 10^{-9}}$$

$$= 7.10 \times 10^{-2} m \ .$$

3. The intensity distribution in the Fraunhofer diffraction pattern for a single slit is given by

$$I(\theta) = I_o \left(\frac{\sin\beta}{\beta}\right)^2 ,$$

where θ is the direction measured from the 'axis', and β is given by $\beta = \frac{\pi}{\lambda} \cdot b\sin\theta$, and b is the slit width. Show that minima occur when $b\sin\theta = m\lambda$, $m = \pm1, \pm2, \dots$.

From the intensity equation $I(\theta) = 0$ when $\sin\beta = 0$ and $\beta \neq 0$. This is satisfied for $\beta = m\pi$, $m = \pm1, \pm2, \dots$ Putting $m\pi$ for β in the equation for β gives

$$m\pi = \frac{\pi}{\lambda} b \sin\theta$$

from which the result follows on dividing through by π and and multiplying through by λ.

Note that although $\sin\beta = 0$ when $\beta = 0$, the function $\frac{\sin\beta}{\beta}$

is equal to 1 when $\beta = 0$, which means a maximum occurs for $\beta = 0$, and hence when $\theta = 0$. This is the central maximum.

4. A beam of polychromatic light ranging from $\lambda = 450$nm to $\lambda = 650$nm falls normally on a transmission grating with grating constant(groove or slit interval) $a = 2.12 \times 10^{-6}$m. The pattern appears on a screen in the focal plane of a converging lens which follows the grating. What focal length must the lens have in order that the second order spectrum is 1.25cm in breadth?

Maxima are given by $a \sin\theta = m\lambda$. m will be 2 for second order maxima. If $y_{2,450}$ and $y_{2,650}$ are the distances of the second maxima from the central maximum for $\lambda = 450$nm and $\lambda = 650$nm, then

$$y_{2,650} - y_{2,450} = 1.25 \times 10^{-2}\text{m}.$$

Now $y_2 = f' \tan\theta_2 \simeq f' \sin\theta_2 = f' \times \dfrac{2\lambda}{a}$, since $\sin\theta_2 = \dfrac{2\lambda}{a}$ when $m=2$.

Hence,
$$y_{2,650} = f' \times \frac{2 \times 650 \times 10^{-9}}{2.12 \times 10^{-6}}$$

and
$$y_{2,450} = f' \times \frac{2 \times 450 \times 10^{-9}}{2.12 \times 10^{-6}} .$$

Thus $y_{2,650} - y_{2,450} = 1.25 \times 10^{-2} = f' \times \dfrac{2(650 - 450) \times 10^{-9}}{2.12 \times 10^{-6}}$

whence $f' = \dfrac{1.25 \times 10^{-2} \times 2.12 \times 10^{-6}}{2 \times 200 \times 10^{-9}} = 6.625 \times 10^{-2}$m.

---

5. What will be the angular separation of the two sodium lines $\lambda = 589.0$nm and $\lambda = 589.6$nm in the first order spectrum produced by a diffraction grating of 500 lines per mm, the light falling normally on the grating?

Using the equation for maxima, $a \sin\theta = m\lambda$, with $m=1$ and λ alternately 589.0 and 589.6nm, we have
$$\sin\theta = \frac{\lambda}{a} = \frac{589.0 \times 10^{-9}}{2 \times 10^{-6}} = 294.5 \times 10^{-3}$$

and $\sin\theta' = \dfrac{\lambda'}{a} = \dfrac{589.6 \times 10^{-9}}{2 \times 10^{-6}} = 294.8 \times 10^{-3}$,

a being $1/500$mm $= 2 \times 10^{-6}$m. Hence, $\theta = 17.13^{\circ}$ and $\theta' = 17.15^{\circ}$ and the angular separation is $\theta' - \theta = 0.02^{\circ}$.

---

6. Parallel light from a hydrogen tube is incident normally
    on a diffraction grating and the angles for the C and F lines
    in the second order spectrum are 41.017° and 29.083°. If the
    wavelength of the C line is λ_c = 656.3nm find (i) the number
    of lines per cm on the grating and (ii) the wavelength for
    the F line.

    (i) We can write for the C line $a \sin\theta = m\lambda_c$,
         whence

    $$a = \frac{m\lambda_c}{\sin\theta} = \frac{2 \times 656.3 \times 10^{-9}}{\sin 41.017°}$$

    $$= \frac{2 \times 656.3 \times 10^{-9}}{0.65629} = 2 \times 10^{-6}\,m$$

    $$= 2 \times 10^{-4}\,cm.$$

    Hence, the number of lines per cm is

    $$\frac{1}{a} = \frac{1}{2 \times 10^{-4}} = 5000 \ .$$

    (ii) Knowing the grating constant, a, we can immediately
         find λ_F :

    $$\lambda_F = a \sin\theta/m = 2 \times 10^{-4} \times \sin 29.083°/2$$

    $$= 4.861 \times 10^{-5}\,cm = 486.1nm.$$

    ---

7. Parallel monochromatic light falls normally on a diffraction
    grating and is focused on a screen by a converging lens which
    follows the grating. How is the Fraunhofer diffraction pattern
    affected by:
          (a) the line spacing, a, on the grating,
     and (b) the number of lines on the grating?

    (a) The mth maximum occurs when $a \sin\theta_m = m\lambda$, where a is
    the grating constant, m is the order, and λ is the wavelength
    of the monochromatic beam; m = 0, ±1, ±2,...

    Hence, the mth order makes an angle θ_m with the normal to the
    grating. If the focal length of the lens is f' the mth maximum
    will be a distance y_m from the central or zeroth maximum, given
    by

    $$y_m = f'\tan\theta_m \simeq f'\sin\theta_m = f'm\lambda/a \ .$$

    The (m+1)th maximum is a distance $y_{m+1} = f'(m+1)\lambda/a$ from
    the zeroth order image.
    Hence, the distance between the mth and (m+1)th maxima is

    $$\Delta y = y_{m+1} - y_m = f'(m+1)\lambda/a - f'm\lambda/a = f'\lambda/a \ .$$

This is independent of the order m so all the maxima are separated by the same distance. Clearly, Δy increases if we increase f' or λ, and decreases with increasing grating constant a.

(b) The intensity distribution in the far-field (Fraunhofer) diffraction pattern for a grating with N lines is given by

$$I(\theta) = I_o \left(\frac{\sin\beta}{\beta}\right)^2 \left(\frac{\sin N\alpha}{\sin\alpha}\right)^2 .$$

I_o is the intensity at a point on the 'grating axis' due to one 'slit' alone, $\beta = \frac{\pi}{\lambda} b \sin\theta$, and $\alpha = \frac{\pi}{\lambda} a \sin\theta$. b is the slit width and a is the grating constant.

Minima occur when $\sin N\alpha / \sin\alpha = 0$;

i.e. when $\alpha = \pm\frac{\pi}{N}, \pm\frac{2\pi}{N}, \ldots, \pm\frac{(N-1)\pi}{N}, \pm\frac{(N+1)\pi}{N}, \ldots$

The omitted values $0, \frac{N\pi}{N}, \frac{2N\pi}{N}$, etc. give maxima.

Hence, a principal maximum extends from $\alpha = -\frac{\pi}{N}$ to $\alpha = +\frac{\pi}{N}$, and so on. Calling this width $2.\Delta\alpha$, we have

$$\Delta\alpha = \frac{\pi}{N} ,$$

which is the half-width of the principal maxima.

But since $\alpha = \frac{\pi}{\lambda} a \sin\theta$, differentiating gives $\Delta\alpha = \frac{\pi}{\lambda} a \cos\theta . \Delta\theta$.

Eliminating $\Delta\alpha$ gives $\frac{\pi}{\lambda} a \cos\theta . \Delta\theta = \frac{\pi}{N}$

or $\Delta\theta = \frac{\lambda}{Na \cos\theta_m}$.

$\Delta\theta$ is the angular subtent of the mth principal maximum half-width measured at the grating. If a converging lens of focal length f' focuses the pattern on a screen in its focal plane, the width of the mth principal maximum will be

$$f'(2.\Delta\theta) = \frac{2 f'\lambda}{Na \cos\theta_m} .$$

Clearly, increasing N reduces the width of each principal maximum. Note that Na is the width of the grating.

8. Suppose a source, illuminating a grating normally, emits light of two wavelengths. Two overlapping Fraunhofer patterns will be formed on a screen when focused by a convex lens. Using Rayleigh's criterion they will just be resolved (separably visible) when the principal maximum of one coincides with the first minimum of the other. Find an expression for the chromatic resolving power of a grating, $\lambda/\Delta\lambda$, where λ is the shorter wavelength and $\Delta\lambda$ is the least resolvable wavelength difference from λ.

The distance from the principal maximum's peak to its first minimum is given by its half-width. Or, in angular terms by $\Delta\theta$ (see last question). This is

$$\Delta\theta = \frac{\lambda}{Na\cos\theta_m} \quad \ldots\ldots\ldots\ldots\ldots(1)$$

Differentiating the grating equation $a\sin\theta = m\lambda$ gives

$$\frac{\Delta\theta}{\Delta\lambda} = \frac{m}{a\cos\theta_m} \quad \ldots\ldots\ldots\ldots\ldots(2)$$

Eliminating $\Delta\theta$ between equations (1) and (2) gives

$$\frac{\lambda}{\Delta\lambda} = mN$$

which is the resolving power of the grating for the mth order.

---

9. What would be the least number of elements in a grating to enable the two components of the sodium D doublet, wavelengths 589.0 and 589.6nm, to be resolved in the second order?

The resolving power of the grating is $\lambda/\Delta\lambda = mN$. Now, $\lambda = 589.0n$ and $\Delta\lambda = 589.6 - 589.0 = 0.6$nm.

Hence,
$$N = \frac{\lambda}{m\cdot\Delta\lambda} = \frac{589.0}{2\times0.6} = 490.8 \ .$$

So 491 lines would be needed to make the second order maxima just visibly separable.

---

10. A telescope objective is 12cm in diameter and has a focal length of 150cm. Light of mean wavelength 550nm from a distant star is imaged by the objective. Calculate the size of the Airy disc.

The Airy disc is the circular disc formed in the image plane (for a point object) and bounded by the first diffraction minimum. For a point distance source the first minimum subtends an angle θ given by

$$\sin\theta = 1.22\frac{\lambda}{d}$$, where λ is the wavelength

and d is the diameter of the circular aperture (the lens in this case). The radius of the Airy disc in the second focal plane of the lens is $f'\tan\theta \simeq f'\sin\theta$, and the diameter is twice this. The diameter is thus

$$2f'\sin\theta = 2f' \times 1.22\lambda/d$$

$$= \frac{2 \times 1.50 \times 1.22 \times 550 \times 10^{-9}}{12 \times 10^{-2}}$$

$$= 16.78 \times 10^{-6}\,\text{m}$$

$$= 0.01678\,\text{mm}.$$

---

11. Light from a distant point source enters a lens of focal length 22.5cm. How large must the lens be if the Airy disc is to be 10^{-6}m in diameter? $\lambda=450$nm.

From question 10 the diameter of the Airy disc, δ, is given by $\delta = 2f'\sin\theta$.

Hence, $$\sin\theta = \frac{\delta}{2f'} = 1.22\frac{\lambda}{d}$$,

and the diameter of the lens must be

$$d = \frac{1.22 \times 2 \times f'\lambda}{\delta} = \frac{1.22 \times 2 \times 22.5 \times 10^{-2} \times 450 \times 10^{-9}}{10^{-6}}$$

$$= 24\,705 \times 10^{-5}\,\text{m}$$

$$= 24.7\,\text{cm}.$$

---

12. An astronomical telescope has an objective of diameter
    150mm and focal length 600mm. What focal length of eye-piece
    would be required to allow the observer to appreciate all
    the detail resolved by the objective? Choose a suitable
    value for the eye's resolution limit ,and use λ=555nm.

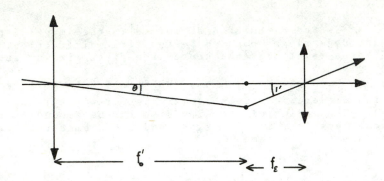

Fig. 15.2

Rayleigh's criterion for the objective states that two point
sources will be resolved in the image if the angular subtent
of the points at the objective is equal to the angular size
of the radius of the Airy disc formed by either image. This
means that the maximum of one Airy disc falls on the first
minimum of the other. If θ is the angular subtent of the two
object points at the objective, we have

$$\theta \simeq \sin\theta = \frac{1.22\lambda}{d} = \frac{1.22 \times 555 \times 10^{-9}}{150 \times 10^{-3}}$$

$$= 4.514 \times 10^{-6} \text{ radians.}$$

The radius of the Airy disc is therefore

$$f_0' \theta = 600 \times 4.514 \times 10^{-6} \text{ mm,}$$

where f_0' is the focal length of the objective. If f_ε' is the
focal length of the eyepiece and we choose 1' for the angular
subtent of the images at the eye, we must have

$$f_0' \theta = f_\varepsilon' \tan 1',$$

whence $f_\varepsilon' = f_0' \theta / \tan 1' = 600 \times 4.514 \times 10^{-6} / 2.909 \times 10^{-4}$

$$= 9.31 \text{mm.}$$

13. Assuming Rayleigh's criterion can be applied to the eye, how far apart must two small lights be to be seen as two at a distance of 1000m? Take the pupil diameter to be 2.5mm, the wavelength to be 555nm, and the eye's refractive index 1.333.

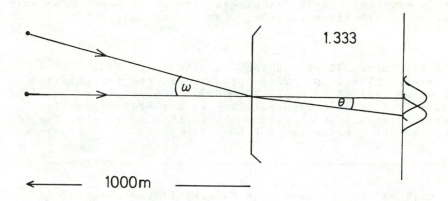

Fig. 15.3

If the lights subtend an angle ω, figure 15.3, and the pupil is assumed in the plane of the cornea, then for small angles and using Snell's law,

$$\omega = 1.333\theta = 1.333(1.22\lambda_e/d)$$

$$= 1.333 \times 1.22 \times (555 \times 10^{-9}/1.333) \div 2.5 \times 10^{-3}$$

$$= 2.71 \times 10^{-4} \text{ rad.,}$$

where $\lambda_e = \lambda/1.333$, the wavelength in the eye.

Hence, the separation of the lights is

$$1000\omega = 2.71 \times 10^{-1}\text{m} = 27.1\text{cm.}$$

All of this assumes a homogeneous atmosphere, a diffraction limited optical system, and a detector-processing system capable of handling the intensity distribution on the retina.* Note that ω = 0.93', which is close to the 1' of arc so-called 'normal' visual acuity.

* See Introduction to Visual Optics, A.H. Tunnacliffe

16. POLARISATION

DICHROISM AND POLAROID

1. Natural light of flux density I is incident on a sheet of
    Polaroid which transmits 32% of the flux. If another sheet
    of this material (HN-32) is placed with its transmission axis
    parallel to the first sheet's, what is the emergent flux
    density?

    HN-32 transmits 32% of the incident flux, I. Since the
    transmitted flux is polarised parallel to the transmission
    axis we can say that $0.64(I/2) = 0.32I$ emerges from the first
    sheet. This is polarised parallel to the second sheet's
    transmission axis which passes 64% of this flux. So, the
    emergent flux density is $0.64(0.32I) = 0.21I$.

    _____

2. What will be the emergent flux density if the analyser is
    rotated 30^0 in the last problem?

    Malus's law states that the emergent flux density $I(\theta)$ is
    given by $I(\theta) = I(0)\cos^2\theta$ when a plane polarised beam of
    flux density $I(0)$ is incident upon an analyser such that the
    angle between the plane of polarisation and the transmission
    axis of the analyser is θ. $I(0)$ is the flux density transmitted
    when $\theta = 0^0$; this is the maximum transmitted, of course.

    Here $\theta = 30^0$ and $I(0) = 0.21I$, so the transmitted flux density
    is
    $$I(\theta) = I(0)\cos^2\theta = 0.21I \cos^2 30^0 = 0.21I \times (\sqrt{3}/2)^2$$

    $$= 0.157I \ .$$

    _____

3. Two perfect linear polarisers (HN-50) are placed with their
    transmission axes vertical and horizontal, respectively. If
    natural light, flux density I, is incident on the first sheet
    a flux density I/2 is transmitted with its plane of polarisation
    vertical. No light emerges from sheet two, of course. A third
    sheet is inserted between the other two with its transmission
    axis at 45^0 to the vertical. What is the emergent flux density
    from the system now?

    I/2 is the flux density incident upon the middle sheet and this
    is plane polarised vertically. Using Malus's law, $\theta = 45^0$ so
    the flux density (irradiance) emerging from the second sheet
    is
    $$\frac{I}{2} \cos^2 45^0 = \frac{I}{2}(\frac{1}{\sqrt{2}})^2 = \frac{I}{4} \ .$$

This strikes the last sheet plane polarised at 45^0 to the vertical (and to the horizontal) and, using Malus's law again, the horizontally polarised emerging irradiance is

$$\frac{I}{4}\cos^2 45^0 = \frac{I}{8} \ .$$

---

4. An unpolarised beam of light of irradiance I passes through two perfect linear polarisers. What must be the relative orientation between their transmission axes for the emergent irradiance to be (a) I/2 , (b) I/4?

(a) Let the angle between the transmission axes be θ. Then, by Malus's law, $I(\theta) = I(0)\cos^2\theta$, for the analyser.

Now, I(0) is the irradiance leaving the polariser and this is I/2, polarised parallel to the polariser's transmission axis. Hence, if $I(\theta) = I/2$ (leaving the analyser) and $I(0) = I/2$ (incident upon the analyser), then

$$\cos^2\theta = \frac{I(\theta)}{I(0)} = \frac{I/2}{I/2} = 1 \ ,$$

whence $\theta = 0^0$, and the transmission axes are parallel.

(b) Here $I(\theta) = I/4$ and $\cos^2\theta = \frac{I(\theta)}{I(0)} = \frac{I/4}{I/2} = \frac{1}{2}$,

whence $\cos\theta = \frac{1}{\sqrt{2}}$ and $\theta = 45^0$. That is, the angle between the transmission axes is 45^0.

POLARISATION BY REFLECTION

5. Find the polarisation angle, i_p, for reflection externally at an air-glass boundary ($n_g = 1.5$). What is the nature of the reflected beam when a parallel beam of natural light is incident on a flat air-glass boundary at $i = i_p$?

Brewster's law states that $\tan i_p = n$, where n is the relative refractive index of the refracting medium. Here n=1.5 and

$$i_p = \arctan n = \arctan 1.5 = 56.3^0 \ .$$

Unpolarised light incident at the polarising angle produces a reflected beam polarised perpendicular to the plane of incidence (i.e. parallel to the surface). Since n is wavelength dependent, a given polarising angle only applies to a given monochromatic incident beam. However, i_p will only vary some 2^0 to 3^0 for light at opposite ends of the visible spectrum which results in most natural light being polarised in the reflected beam when it is incident about a mean i_p value.

---

6. Fresnel's intensity reflectances for reflection at a dielectric (non-conducting) material are

$$R_{\shortparallel} = \frac{I_{r\shortparallel}}{I_{i\shortparallel}} = \frac{\tan^2(i-i')}{\tan^2(i+i')}$$

and

$$R_{\perp} = \frac{I_{r\perp}}{I_{i\perp}} = \frac{\sin^2(i-i')}{\sin^2(i+i')} \ ,$$

where $\shortparallel$ and $\perp$ refer to the electric vector components parallel to and perpendicular to the plane of incidence, $I_{i\shortparallel}$ is the incident intensity (I_i) parallel to the same plane, and the subscript r refers to reflected intensities.

Use Fresnel's equations to show that it is possible for the reflected light to be entirely polarised parallel to the surface (perpendicular to the plane of incidence) and, further, use Snell's law in addition to derive Brewster's law, $\tan i_p = n'/n$, where i_p is the polarisation angle and n and n' are the absolute refractive indices of the object and image space media. Monochromatic light is assumed!

Note that $R_{\shortparallel} = 0$ when $\tan(i+i') = \infty$; that is, when $i+i' = 90^0$. Under this condition $R_{\perp}$ is non-zero so the reflected light is entirely polarised perpendicular to the incident plane (which is therefore parallel to the surface). When this occurs $i=i_p$, by definition.

Snell's law states that $n \sin i_p = n' \sin i'$; but $i+i' = i_p+i' = 90^0$, from the last paragraph, whence $i' = 90^0 - i_p$.
Rewriting Snell's law, substituting for i', gives

$$n \sin i_p = n' \sin i' = n' \sin(90^0 - i_p) = n' \cos i_p \ .$$

It follows that

$$\frac{\sin i_p}{\cos i_p} = \tan i_p = \frac{n'}{n} \ , \text{ which is Brewster's law.}$$

(We have used $\sin(90^0 - x) = \cos x$, and $\frac{\sin x}{\cos x} = \tan x$).

---

7. A beam of light is incident on an air-water surface (n_w=1.333) at 30^0. Find the degree of polarisation of the reflected beam. The degree of polarisation can be shown to be given by $(R_\perp - R_{\shortparallel})/(R_\perp + R_{\shortparallel})$.

We can use Fresnel's equations to find $R_\perp$ and $R_{\shortparallel}$, but firstly we need to know the angle of refraction i'. Using Snell's law,

$$\sin i' = \frac{n}{n'}\sin i = \frac{1}{1.333}\sin 30^0 = 0.3751,$$

whence i' = 22.03^0.

So, $$R_\perp = \frac{\sin^2(i-i')}{\sin^2(i+i')} = \frac{\sin^2(30^0 - 22.03^0)}{\sin^2(30^0 + 22.03^0)} = 0.0309$$

and $$R_{\shortparallel} = \frac{\tan^2(i-i')}{\tan^2(i+i')} = \frac{\tan^2(30^0 - 22.03^0)}{\tan^2(30^0 + 22.03^0)} = 0.0119 .$$

The degree of polarisation is therefore

$$\frac{R_\perp - R_{\shortparallel}}{R_\perp + R_{\shortparallel}} = \frac{0.0309 - 0.0119}{0.0309 + 0.0119} = 0.4439 = 44.39\%$$

---

8. Unpolarised mercury light (λ=546.1nm) is incident at an angle of 58.02^0 on a glass plate. If the reflected beam is found to be completely polarised parallel to the surface, find the refractive index of the glass.

Using Brewster's law $\tan i_p = n_g$. Now, $i_p = 58.02^0$,
so $n_g = \tan 58.02^0 = 1.602$.

---

9. A glass plate (n_g=1.673) is immersed in water (n_w=1.333). Find the polarisation angles for both internal and external reflection at the water-glass boundary.

External reflection

$$i_p = \arctan\frac{n_g}{n_w} = \arctan\frac{1.673}{1.333} = 51.45^0 .$$

Internal reflection

$$i_p = \arctan\frac{n_w}{n_g} = \arctan\frac{1.333}{1.673} = 38.55^0.$$

Note that the two angles add to 90^0 which can be shown to
be generally true.

BIREFRINGENCE

10. Calculate the angle between the o- and e-rays emerging from
a calcite Wollaston prism of wedge angle 15^0, figure 16.2,
($n_b=1.66$ and $n_e=1.49$).   State whether calcite is negatively
or positively birefringent.

Birefringence is defined as $\Delta n = n_e - n_b$, so calcite is negatively
birefringent since $\Delta n = -0.17$.   The values of n_e and n_b are
quoted for $\lambda=589.3$nm.

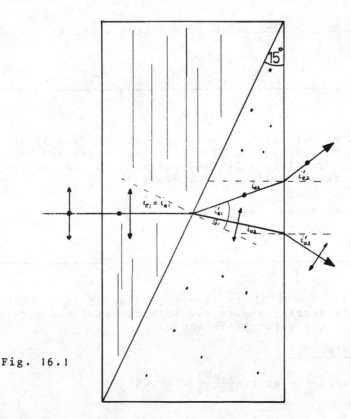

Fig. 16.1

The unpolarised light is incident normally on the first face and proceeds unrefracted to the diagonal interface where the angles of incidence of the e- and o-rays are $i_{e1} = i_{o1} = 15^0$.

The e-ray, shown with its electric vector, $\updownarrow$, parallel to the optic axis in the first prism segment, becomes an O-ray in the second prism segment where its vector is then perpendicular to the optic axis. Applying Snell's law at the interface

$$n_e \sin i_{e1} = n_o \sin i_{o1}'$$

whence $\quad \sin i_{o1}' = \dfrac{n_e}{n_o} \sin i_{e1} = \dfrac{1.49}{1.66} \sin 15^0 = 0.232$

and $\quad i_{o1}' = 13.4^0$.

Similarly, the o-ray in the first prism becomes an e-ray in the second prism.
Hence,

$$n_o \sin i_{o1} = n_e \sin i_{e1}' \, ,$$

or $\quad \sin i_{e1}' = \dfrac{n_o}{n_e} \sin i_{o1} = \dfrac{1.66}{1.49} \sin 15^0 = 0.288,$

and $\quad i_{e1}' = 16.7^0$

Clearly, the e-ray in the second prism bends upwards and the o-ray downwards.
Considering refraction at the last surface,

$$1 \times \sin i_{e2}' = n_e \sin i_{e2} = 1.49 \sin 1.7^0 = 0.044,$$

and $\quad i_{e2}' = 2.52^0$.

Also, $\quad 1 \times \sin i_{o2}' = n_o \sin i_{o2} = 1.66 \sin 1.6^0 = 0.046,$

and $\quad i_{o2}' = 2.64^0$.

The angle between the emergent rays is

$$i_{e2}' + i_{o2}' = 2.52^0 + 2.64^0 = 5.16^0 \, .$$

11. Calculate the angle between the o- and e-rays emerging from
the quartz Wollaston polarising beamsplitter in figure 16.2 .
n_o = 1.5443 and n_e =1.5534 .

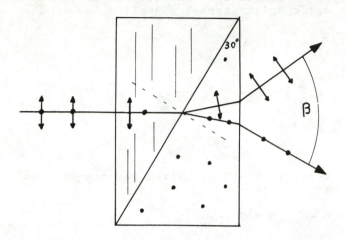

Fig. 16.2

The wedge angle is 30^0 and the optic axis in the first prism
segment lies in the plane of the figure and parallel to the
first surface. The optic axis in the second prism segment is
perpendicular to the paper.

Unpolarised light is incident on the first face. Since it
strikes normally there is no refraction here. In this segment
there is a phase difference, the e-wave travelling more slowly
than the o-wave since $n_e > n_o$. This is shown in the figure
where the e-vector, parallel to the optic axis, $\updownarrow$, is behind
the o-vector, • , which is perpendicular to the optic axis.

At the diagonal interface Snell's law applies. Here the e-wave
in the first segment becomes an o-wave in the second segment
since the vectors which were parallel to the optic axis before
the interface are perpendicular to the optic axis in the second
segment.

(i) Denoting the e-ray angle of incidence at the interface
by i_{el} and the refracted ray (which becomes an o-ray) as i'_{ol} ,
then

$$n_e \sin i_{el} = n_o \sin i'_{ol} \, ,$$

whence $\sin i'_{ol} = \dfrac{n_e}{n_o} \sin i_{el} = \dfrac{1.5534}{1.5443} \sin 30^0 = 0.5029$.

Thus, $i'_{ol} = 30.19^0$. This ray is bending upwards in the figure.

Consideration of the triangles involved shows this ray is incident on the last face at $i_{o2} = 0.19^{0}$. It emerges making an angle i'_{o2} given by

$$1 \times \sin i'_{o2} = n_o \sin i_{o2} = 1.5443 \sin 0.19^{0},$$

whence $i'_{o2} = 0.293^{0}$.

(ii) Considering the o-ray in the first segment the angle of refraction (i'_{e1}) at the interface, where it becomes an e-ray, is given by

$$n_o \sin i_{o1} = n_e \sin i'_{e1}.$$

Hence, $\sin i'_{e1} = \dfrac{n_o}{n_e} \sin i_{o1} = \dfrac{1.5443}{1.5534} \sin 30^{0} = 0.4971$,

and $i'_{e1} = 29.80^{0}$.

This ray is bending downwards in the figure. It makes an incident angle at the last face $i_{e2} = 0.2^{0}$ and the emergent angle i'_{e2} is given by

$$1 \times \sin i'_{e2} = n_e \sin i_{e2} = 1.5534 \sin 0.2^{0} = 0.005422.$$

Hence, $i'_{e2} = 0.311^{0}$.

Now, one ray is bending upwards and the other downwards, so the angle between them is $0.293^{0} + 0.311^{0} = 0.604^{0}$. The angles in the figure have been exaggerated for clarity.

12. Describe the properties of a half-wave retarder or retardation
    plate. A piece of sellotape is stretched slightly along its
    length and stuck to a microscope slide. It is placed between
    crossed polaroids at 45^{0} to the transmission axes whereupon
    light passes through the system over the area covered by the
    sellotape. Explain.

A uniaxial birefringent crystal cut to form a parallel sided
plate with its optic axis parallel to the front surface acts
as a retardation plate. If its thickness is t, the optical
path difference for the o- and e-waves is

$$O.P.D. = (|n_o - n_e|)t$$

where $n_o t$ and $n_e t$ are the optical path lengths for the
ordinary and the extraordinary waves, respectively. When the
O.P.D. equals an odd number of half-wavelengths the two rays
will be a half-wavelength out of phase on emerging from the
plate. This can be stated mathematically as

$$(|n_o - n_e|)t = (2m+1)\frac{\lambda}{2} \quad , \quad m = 0, 1, 2,....$$

Such a plate is a half-wave plate and is clearly wavelength
dependent. The thickness t depends on the value of m, the
minimum thickness occurring when m=0 for a given λ, and given
n_o and n_e. Figure 16.3 shows the effect on a plane polarised
wave.

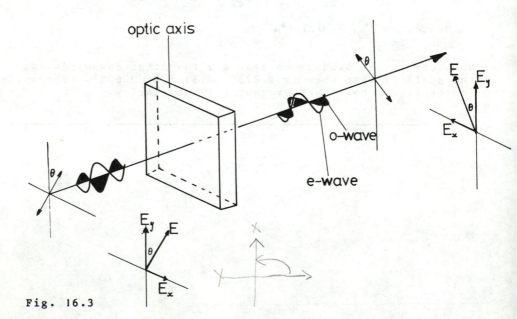

Fig. 16.3

If the amplitude (the electric vector) is E it can be considered as horizontal and vertical components E_x and E_y. The E_y component is parallel to the optic axis and is therefore an e-wave, whilst the E_x component perpendicular to the optic axis is an o-wave. Because each wave experiences a different refractive index they are transmitted at different speeds and leave the plate with a phase change governed by the plate thickness, the birefringence, and the wavelength. The emergent resultant E is 'flipped' through 2θ degrees for a half-wave plate.

Sellotape is birefringent and can be made more so by stretching it a little. When a single piece of sellotape is stuck on a microscope slide and held at 45^0 between crossed polaroids white light emerges from the system over the area covered by the sellotape (cellophane tape). Since the O.P.D. is given by $(|n_o - n_e|)t$ and the refractive indices are wavelength dependent, it is possible that more than one wavelength will be flipped over by $2 \times 45^0 = 90^0$ and emerge from the analyser.

Wavelengths for which $(|n_o - n_e|)t$ is a full wave-length will emerge in phase and will not suffer any 'rotation'. They will be absorbed by the analyser. Other wavelengths will be in various states of elliptical polarisation and their components parallel to the transmission axis of the analyser will emerge. If the combination of emergent wavelengths is 'just right' the eye will detect 'white' light.

It is interesting to build up layered steps of stretched sellotape on the slide when different thicknesses will present different colours to the observer.

---

13. Find the minimum thickness of a quartz retarder for it to act as a half-wave plate for light of 590nm. ($n_e = 1.5534$, $n_o = 1.5443$)

From question 12 we know that $t = \dfrac{(2m+1)\lambda/2}{(|n_o - n_e|)}$ for a half-wave

plate, and t is a minimum when m=0.

Thus, $t = \dfrac{\lambda/2}{|n_o - n_e|} = \dfrac{590/2}{|1.5443 - 1.5534|} = 32\,418$nm

---

- 144 -

14. Two polaroids are placed with their transmission axes parallel
and placed between these is a piece of sellotape stuck to a
microscope slide. If the 'length' of the sellotape is placed
at 45° to the transmission axes no light emerges from the system
over the area covered by the sellotape. Explain.

Since no light emerges the half-wave plate condition (see last
question) must be satisfied for all wavelengths. If t is
constant this means $|n_o - n_e|$ must just vary with the wave-
length sufficiently to make the sellotape a half-wave plate
for all wavelengths.

---

15. A crystalline positive uniaxial prism, figure 16.4, produces
minimum deviation angles of 46° and 40°. Determine n_e and n_o.

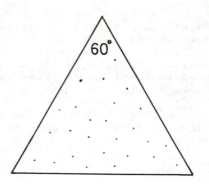

Fig. 16.4

For any ray entering the prism in the plane of the paper it
will have electric vector components parallel and perpendicular
to the optic axis (shown perpendicular to the paper). Since
it is positively birefringent $n_e > n_o$. Hence, the minimum
deviation angle 46° is produced by the e-ray, and

$$n_e = \frac{\sin((A+D)/2)}{\sin A/2} = \frac{\sin(30°+23°)}{\sin 30°} = 1.597$$

and

$$n_o = \frac{\sin(30°+20°)}{\sin 30°} = 1.532 .$$

16. Describe the phenomenon known as stress birefringence or
    photoelasticity. What is its significance in ophthalmic
    practice?

Normally transparent isotropic substances can be made optically
anisotropic by application of mechanical stress. Under
compression or tension the material takes on properties of a
negative or positive uniaxial crystal, respectively. The optic
axis is in the direction of the stress and the birefringence,
$n_e - n_o$, is proportional to the stress. If the stress is not
uniform over the object neither will be the birefringence.

Ophthalmic lenses are subjected to stress when mounted in
spectacle frames. Since the stress varies from point to point
and the thickness is not uniform, the O.P.D. = $(|n_o - n_e|)t$
also varies from point to point. Under white light illumination
a stressed lens placed between crossed polaroids exhibits
ISOCHROMATIC regions for which the O.P.D. is constant and each
region corresponds to a particular colour. In addition, there
are black bands known as ISOCLINICS which correspond to regions
where the direction of the stress is parallel to the polariser's
transmission axis. In this case the electric vector cannot be
resolved into mutually orthogonal components along and perpendicular to the local optic axis; the wave passes through the
lens unaffected and is absorbed by the analyser. This latter
effect will be noticed when the length of the sellotape in
question 14 is placed parallel to or perpendicular to the
transmission axis of the polariser: no light passes through
the analyser and the sellotape appears black.

17.   STOPS

1. Define the terms aperture stop, entrance pupil, and exit
    pupil. Two thin positive lenses Λ_1 and Λ_2 are separated by
    5cm. Their diameters are 6cm and 4cm respectively, and
    their focal lengths are $f_1' = +9$cm and $f_2' = +3$cm. A 1cm diameter
    stop is located between them and 2cm from Λ_2. Find (a) the
    aperture stop (A.S.) and (b) the positions and sizes of the
    entrance and exit pupils for an axial point object, B, 12cm
    in front of Λ_1.

Definitions

The aperture stop is that aperture (lens, stop proper, etc.)
which limits the cone of rays from a given axial object point.
It therefore limits the amount of light reaching the image and
thus controls the latter's illuminance.

The entrance pupil is the image of the aperture stop formed by
the optical components in front of it. Where there are no
components in front of the aperture stop the aperture stop
is also the entrance pupil.

The exit pupil is the image of the aperture stop formed by
those components following it. Again, where there are no
components following it the aperture stop is also the exit
pupil.

The entrance and exit pupils are conjugate with respect to the
system as a whole. That is, a ray directed towards the centre
of the entrance pupil, say, passes through the centre of the
A.S., and leaves the system as though from the centre of the
exit pupil.

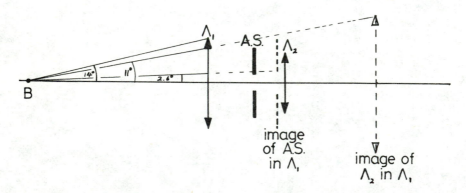

Fig. 17.1

To find the A.S. we must find which stop, or stop image,
subtends the smallest angle at the axial object point B.

(i) Λ_1 subtends a half-angle ϕ_1 at B given by $\tan\phi_1 = 3/12 = 0.25$, whence $\phi_1 = 14^0$.

(ii)   Considering the stop, we must find its image formed by Λ_1.   The stop is 3cm from Λ_1 so the image of the stop is ℓ' from Λ_1, given by

$$\frac{1}{\ell'} = \frac{1}{f_1} + \frac{1}{\ell} = \frac{1}{9} + \frac{1}{-3} = -\frac{2}{9} ,$$

whence $\ell' = -4.5$cm.   The image is on the same side of Λ_1 as the stop.   The magnification is

$$m = \frac{\ell'}{\ell} = \frac{-4.5}{-3} = 1.5,$$

so the radius of the stop image is

$$1.5 \times \text{stop's radius} = 1.5 \times 0.5 = 0.75\text{cm}.$$

This image is $12 + 4.5 = 16.5$cm from B, and the half-angular subtent is given by $\tan\phi_2 = 0.75/16.5 = 0.04545$, whence $\phi_2 = 2.6^0$.

(iii)   We must now consider the image of Λ_2 formed by the optical components to the left of it.   There is only one component, Λ_1, so we consider the image of Λ_2 in Λ_1. Now, Λ_2 is 5cm from Λ_1 so its image distance is given by

$$\frac{1}{\ell'} = \frac{1}{f_1} + \frac{1}{\ell} = \frac{1}{9} + \frac{1}{-5} = \frac{5-9}{45} = -\frac{4}{45} ,$$

whence $\ell' = -\frac{45}{4} = -11.25$cm.

Thus, the image of Λ_2 is 45/4cm from Λ_1 and on the same side as Λ_2.   The magnification is

$$m = \frac{\ell'}{\ell} = \frac{-45/4}{-5} = \frac{9}{4} = 2.25 .$$

Hence, the half-angular subtent of the image of Λ_2 at B is given by

$$\tan\phi_3 = \frac{\text{radius of image}}{\text{distance of image from B}}$$

$$= \frac{m \times 2}{12 + 11.25}$$

$$= \frac{2.25 \times 2}{23.25} = 0.1935 .$$

So $\phi_3 = 11^0$.

Clearly, ϕ_2 is the smallest angle, so the image of the stop is the entrance pupil and the stop is the A.S.

Knowing which element is the A.S. we find the exit pupil by finding the image of the A.S. in the components behind (to the right of) it. Here Λ_2 is the only image forming component behind the A.S. and the object distance is the distance separating them, 2cm. Its image position is given by

$$\frac{1}{\ell'} = \frac{1}{f'} + \frac{1}{\ell} = \frac{1}{3} + \frac{1}{-2} = \frac{2-3}{6} = -\frac{1}{6} ,$$

so $\ell' = -6$cm. The magnification is m = ℓ'/ℓ = (−6)/(−2) = 3, so the diameter of the exit pupil is

$$3 \times \text{diameter of A.S.} = 3 \times 1 = 3\text{cm}.$$

The exit pupil is on the same side of Λ_2 as is the A.S. These details are shown in figure 17.1 .

---

2. A thin lens with diameter 5cm and focal length +4cm has a 3cm diameter stop located 2cm in front of it. An object 1.5cm high stands on the axis 9cm in front of the lens. Locate graphically and by formula (a) the position, and (b) the size of the exit pupil. (c) Locate the image of the object graphically by drawing two marginal rays and the chief ray from the top of the object. Define the term chief ray.

We will define the term chief ray firstly. From a stated point on the object that ray which passes through the centre of the aperture stop is the chief ray for that object point. Clearly, all object points have a chief ray (provided the object lies within the object space field of view of the instrument, of course). A chief ray is found by directing a ray, from the object point in question, towards the centre of the entrance pupil. It then passes through the centre of the A.S. after refraction by those elements in front of the latter.

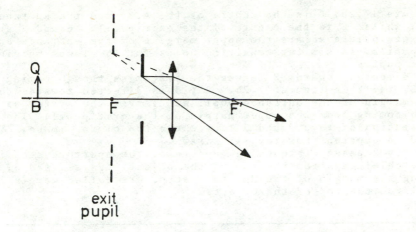

exit
pupil

Fig. 17.2a

The above figure is drawn to scale ($0.5\text{cm} \equiv 1\text{cm}$). Clearly,
the stop subtends a smaller angle at B than the lens rim does.
Hence, the stop is the A.S. Since there is no optical element
in front of it it is also the entrance pupil. Its image in
the lens is the exit pupil and the latter is shown located
graphically in figure 17.2a

Location of the exit pupil position and size by formula

The stop is 2cm to the right of the first focal point, F, of the
lens. Using Newton's equation

$$x' = ff'/x = (-4)(4)/2 = -8\text{cm},$$

placing the image 8cm to the left of F', figure 17.2a. That is,
at F.

The magnification is $m = -\dfrac{f}{x} = \dfrac{-(-4)}{+2} = +2$, so the exit pupil

diameter is $2 \times$ stop diameter, or 6cm.

Location of the image

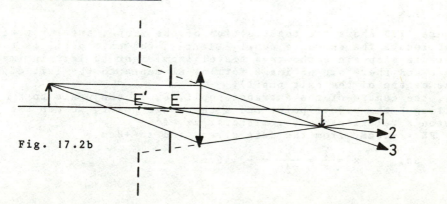

Fig. 17.2b

In figure 17.2b, E is the centre of the A.S. and the entrance
pupil, whilst E' is the centre of the exit pupil. E and E' are
conjugate points as are the upper margins of the entrance and
exit pupils. A similar condition holds for the lower margins
of the pupils.
A ray directed towards E leaves the system as though coming
from E', ray 2 in figure 17.2b. Ray 3 is directed towards the
upper margin of the entrance pupil and leaves the system as
though coming from the upper margin of the exit pupil. The
intersection of rays 2 and 3 locate the top of the image. It
should be apparent how ray 1 is drawn.
Since ray 2 passes through the centre of the aperture stop it
is the chief ray from the top of the object. Rays 1 and 3,
grazing the margin of the aperture stop, are called marginal
rays (from the top of the object).

---

3. An exit pupil, 3cm in diameter, is located 10cm in front of
   a spherical mirror of radius +16cm. An object 2cm high is
   placed on the axis 6cm in front of the mirror. Find the
   entrance pupil's position and size, and locate the image by
   drawing a marginal ray and the chief ray (from the top of the
   object).

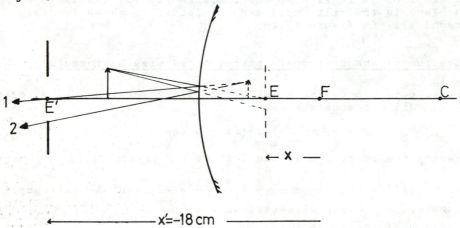

Fig. 17.3

Figure 17.3 shows the construction of the image, but first we
must locate the entrance pupil. Clearly, the exit pupil is the
aperture stop since the rays travel from right to left in image
space and there are no image forming elements to the left of E'
(the centre of the exit pupil).
Now, the centre of the entrance pupil, E, is conjugate to E'.
Calling E the object point and E' the image point, we can use
Newton's relationship for mirrors, $xx' = f^2$.
$x' = FE' = -18cm$, from the data given, and $f = +8cm$.

So, $$x = \frac{f^2}{x'} = \frac{8^2}{-18} = -3.56cm.$$

That is, E is 3.56cm to the left of F. To find the size of
the entrance pupil we first find the magnification;
$$m = -\frac{f}{x} = -\frac{8}{(-3.56)} = +2.28 \ .$$

Remember we are using the entrance pupil as a virtual object
for the real image, the exit pupil.

Now, $\frac{h'}{h} = 2.28$, so h = h'/2.28 = 3/2.28 = 1.32cm.

The figure is drawn to scale with 0.5cm ≡ 1cm.
Ray 1 directed towards the centre of the entrance pupil, E,
is reflected through the centre of the exit pupil, E'. Ray 2,
directed towards the lower edge of the entrance pupil, is
reflected through the lower edge of the exit pupil. Producing
these rays backwards locates the top of the image at their
intersection. Note that the conventional construction ray
from the top of the object towards C meets the top of the
image, as indeed it should!

---

4. Define the terms field stop, entrance port, and exit port.
Construct a scale diagram of the object and image angular
fields of view* for a lens with aperture 4cm and $f_1' = +6$cm,
used as a magnifier. Assume the eye's pupil is 1cm in diameter
and is located at the reduced eye's surface. The distance
between the lens and the eye is 3.5cm. If a 1cm high object
is located on the axis 5cm from the lens, find the image size
and position. How much ocular accommodation would be required
to focus the image by an eye with ocular refraction -1.00D?

Definitions

The field stop, which may be a stop proper or the rim of an
optical element in the system, limits the angle which rays
can make with the axis. It therefore limits the extent of
the object (the field) which can be imaged in the focal plane.

The image of the field stop formed by that part of the system
in front of it is called the entrance port. The exit port is
the image of the field stop formed by that part of the system
behind the field stop.
The entrance port and the exit port are conjugate with respect
to the optical system as a whole.

Determination of the field stop in a system

That stop or stop image subtending the smallest angle at the
centre of the entrance pupil, E, is the entrance port. If
the latter is an image then its object is the field stop.

* Unless otherwise stated assume this refers to the field of
  half illumination.

In the situation we have here we can assume the eye pupil
is the aperture stop. It is easily verified that the image
of the pupil is 8.4cm to the right of the lens, virtual,
erect, and 2.4cm in diameter, and this image subtends a smaller
angle at B than the lens itself does. This image is the
entrance pupil, E in figure 17.4 . The pupil itself is the
exit pupil since there is no optical component behind it.
The only other candidate as the field stop is the lens rim.
Since there is no optical system in front of the lens it is
both the field stop and the entrance port. The exit port is
the image of the lens (the field stop) in the system behind
it (the reduced eye's cornea).
Assuming axial myopia the cornea is +60D. The image of the
lens in this single surface is found in the usual way, as
follows:

The lens is -3.5cm from the surface. Writing the corneal
power F_e = +60D, the object vergence is

$$L = \frac{n}{\ell} = \frac{1}{-0.035} = -28.57D.$$

Then $\qquad L' = L + F = -28.57 + 60 = +31.43D,$

whence $\qquad \ell' = \frac{n'}{L'} = \frac{1.333}{31.43} = +0.04241m = +4.241cm,$

where n'= 1.333 is the refractive index of the eye.

The image size is

$$h' = mh = \frac{L}{L'}h = \frac{-28.57}{+31.43} \times 4 = -3.64cm,$$

the figure 4 being the diameter of the lens, so h'= -3.64cm
is the diameter of the exit port and the negative sign
indicates it is inverted.
Figure 17.4 shows the stops, pupils, and ports drawn to scale.

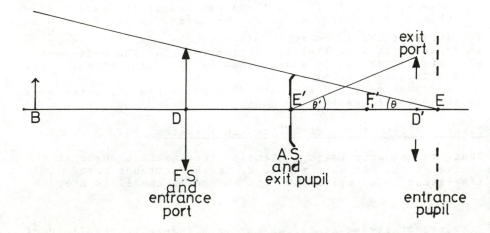

Fig. 17.4

The angular field of view in the object space, 2θ, is the angular subtent of the entrance port at the centre of the entrance pupil. The angular field of view in the image space, $2\theta'$, is the angular subtent of the exit port at the centre of the exit pupil.

In this case,

$$\theta = \arctan(2/8.4) = 13.4^0,$$

and $$\theta' = \arctan(1.82/4.24) = 23.2^0$$

Calculation of image position in the lens

The object is 5cm from the lens so $L = \dfrac{n}{\ell} = \dfrac{1}{-0.05} = -20D,$

and the lens power is +16.67D.
Hence, $L' = L + F = -20 + 16.67 = -3.33D$

and $\ell' = \dfrac{n'}{L'} = \dfrac{1}{-3.33} = -0.30m = -30cm.$

This makes the image $(30 + 3.5) = 33.5cm$ from the eye. Since this is nearly $\frac{1}{3}m$ the vergence arriving at the eye will be very nearly $-3D$ ($-2.99D$ actually), so the eye being $-1.00D$ myopic will need to accommodate 2 dioptres.

The image size is $mh = \dfrac{L}{L'}.h = \dfrac{-20}{-3.33} \times 1 = 6cm.$

---

5. The focal length of a thin positive lens of 4cm diameter is 12cm. If this lens is placed midway between the eye and a large object 10cm from the eye, what width of the object can be seen through the lens?

Firstly, we assume the question refers to the object field of half-illumination, especially since no pupil size has been given. Assuming the reduced eye and pupil relationship of question 4, the eye pupil will be the exit pupil and its centre E' is conjugate with the centre of the entrance pupil E. Now, E' (the eye pupil and exit pupil) is 5cm from the lens and if this is regarded as the image distance then E is the object at a distance $AE = \ell$ from the lens, given by

$$\frac{1}{\ell} = \frac{1}{\ell'} - \frac{1}{f'} = \frac{1}{5} - \frac{1}{12} = \frac{7}{60} ;$$

i.e. $\ell = +8.6cm.$

The data are shown in figure 17.5

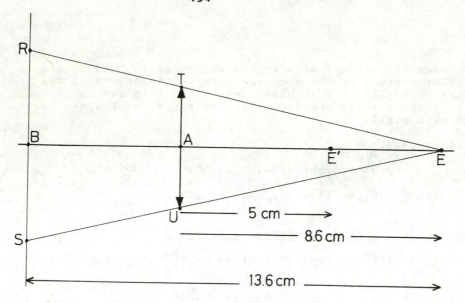

Fig. 17.5

The lens is the entrance port, so the object field of view
is the angle TÊU. The extent of the object visible is RS,
and from similar triangles we have

$$RS = BE \cdot \frac{TU}{AE} = 13.6 \times \frac{4}{8.6} = 6.3 cm.$$

6. In the example of question 5, if the pupil is 0.6cm in diameter
over what extent of object is there no diminution of illuminance
(illumination)?

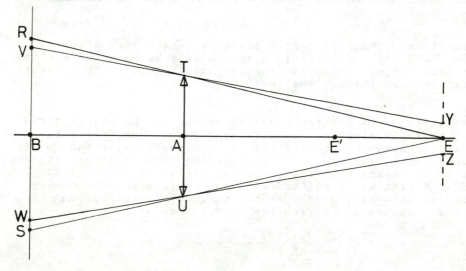

Fig. 17.6

Here we need to know the diameter of the entrance pupil at
E. Denoting the eye pupil diameter by h' and the entrance
pupil diameter by h we have, from the magnification formula,

$$h = h'.\frac{\ell}{\ell'} = 0.6 \times \frac{8.6}{5} \simeq 1\text{cm} .$$

The entrance pupil is shown as YZ in figure 17.6. Since the
lens, TU, is the field stop and the entrance port, the lines
YTV and ZUW delineate that part of the object which suffers
no fall-off in observed illuminance. VW = 5.7cm is the extent
of the object seen fully illuminated and this has been taken
directly from the figure which is drawn to scale. RS = 6.3cm
is also drawn from the data in question 5.

---

7. Determine the fully illuminated field of view (image space)
    for an astronomical telescope for which the data are:

>   magnifying power = 20
>   objective: $f_o' = 60$cm, diameter = 5cm
>   eye-lens: diameter = 1cm
>   eye-pupil diameter = 3mm.
Show where the eye pupil should be located.

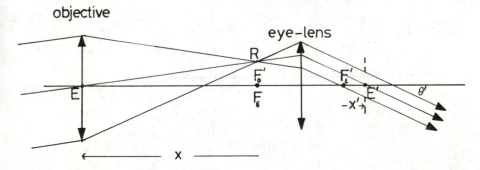

Fig. 17.7

We can safely assume the objective is the aperture stop and,
since there is no refracting element in front of it, it is
also the entrance pupil. Its image in the eye-lens is the
exit pupil the centre of which is shown at E', figure 17.7.
Since we are dealing with image space field of view we shall
need to know the position and size of the exit pupil.
Using Newton's formula then to find the image of the objective
in the eye-lens, we have

$$x' = f_\varepsilon f_\varepsilon'/x = (-3)(3)/(-60) = +0.15\text{cm},$$

where the eye-lens focal length, f_ε', was obtained using the
magnifying power relationship $M = f_o'/f_\varepsilon$.

This places the exit pupil 0.15cm to the right of F'_ϵ, the second focal point to the eye-lens. It is a real inverted image. The diameter of the exit pupil is given by

$$h' = mh = -\frac{f\epsilon}{x}\cdot h = -\frac{(-3)}{(-60)}\times 5 = -0.25\text{cm}.$$

This is less than the eye-pupil diameter so if the latter is placed at E' all the light entering the objective will pass into the eye.

The eye lens is the field stop for the telescope and, since there is no optical component following it, it is also the exit port. The semi-angular field of view for full illuminance (in image space) is given by θ' in figure 17.7.

Notice that any increase in obliquity of the rays incident on the objective will cause some rays to miss the eye-lens. We can find θ' more clearly if we re-draw part of figure 17.7 as in figure 17.8. Since the vertical diameter of the bundle of rays is equal to the diameter of the exit pupil (2.5mm), and the radius of the eye-lens is 5mm, we can easily see that $\theta' = \arctan(3.75/31.5) = 6.79^0$.

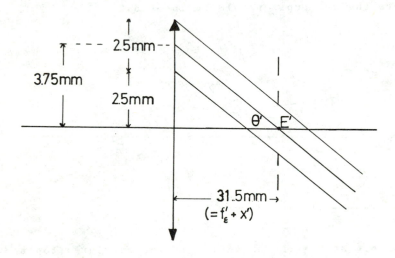

        Fig. 17.8

(In practice a field stop proper would be placed in the common focal plane of the objective and the eye-lens and it would have a radius $F'_o R$, see figure 17.7, so that rays focusing above R would be entirely cut off. This results in a sharply outlined field of view with full illumination.

8.(i) With reference to the photographic camera explain
    (a) the purpose of the adjustable aperture, and (b) the
    meanings of depth of focus and depth of field.
  (ii) Define the term exposure. If the correct exposure time
    with an aperture of f/8 is 1/30 second, what would be the
    exposure time with the f/16 aperture for the exposures to be
    the same?
  (iii) If the depth of focus with the f/8 aperture is 0.04mm,
    what would be the depth of focus with the f/16 aperture?

(i) (a) The adjustable aperture serves two purposes. One is
to control the intensity of the light in the image plane, the
intensity being proportional to the area of the aperture.
Since the area of the aperture is $\pi \times (DIAMETER/2)^2$, the intensity
is proportional to the square of the diameter.
Secondly, the diameter of the aperture affects the depth of
focus.

    (b) To appreciate the meanings of depth of focus and depth
of field refer to figure 17.9a and b.

Fig. 17.9

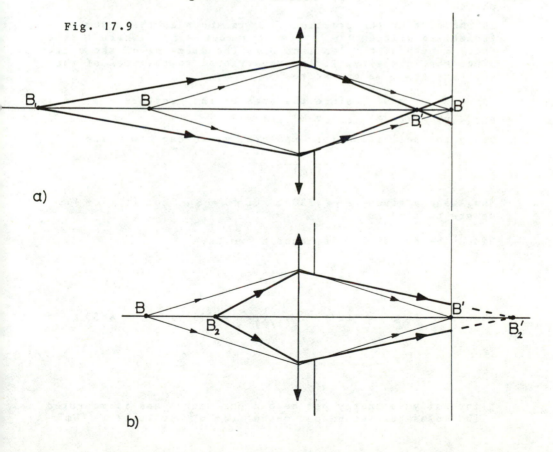

a)

b)

The lens, the aperture stop, and the screen should be apparent.
B is an object focused at B' on the screen. In 17.9a an object,
B_1, further from the lens than B, produces a blur disc on the
screen. Similarly, B_2, nearer the lens than B, produces the
same size blur disc on the screen. Generally, for photographic
purposes blur discs f'/1000 in diameter are subjectively
perceived as 'focused' points. Given that the blur discs in
the figure have this limiting size, then objects from B_1 to B_2
will produce 'clear' images. The distance $B_1 B_2$ is the depth
of field whilst the related image space distance $B_1' B_2'$ is known
as the depth of focus.

(ii) If I is the intensity* in the image plane and Δt is the
length of time during which the camera shutter is open, then
the exposure, E, is given by

$$E = I.\Delta t \ .$$

Δt is known as the exposure time. If we look at the units of
$I.\Delta t$ we have $Js^{-1}m^{-2}s = Jm^{-2}$, which tells us that exposure is
the measure of the incident energy per metre^2 in the image
plane.

In photography the fraction f'/D is known as the f-number
(sometimes written f/#); i.e. f-number = f'/D, where f' is the
focal length of the lens and D is the diameter of the aperture.
Since the intensity, I, is proportional to the area of the
aperture stop, we can write

$$I \propto D^2 \ , \text{ since the area of the aperture } = \pi(\tfrac{D}{2})^2 \ .$$

But, $D = f'/(f/\#)$, so $I \propto 1/(f/\#)^2$.

Using the definition for exposure, E, we can now write

$$E \propto \frac{\Delta t}{(f/\#)^2} \ .$$

Now, we are given $\Delta t_1 = 1/30s$, $(f/\#)_1 = 8$, and $(f/\#)_2 = 16$;
we are to find Δt_2.

Since the exposure E is constant we have

$$E \propto \frac{\Delta t_1}{(f/\#)_1^2} = \frac{\Delta t_2}{(f/\#)_2^2} \ ,$$

whence $\Delta t_2 = (f/\#)_2^2 \cdot \dfrac{\Delta t_1}{(f/\#)_1^2} = (16)^2 \times \dfrac{\tfrac{1}{30}}{8^2} = 4/30 \text{ s} .$

* Intensity is energy per second per metre^2 and is measured
  in Joules per second per metre^2 ($Js^{-1}m^{-2}$ or Watts m^{-2} (Wm^{-2})).

(iii) The depth of focus is approximately proportional to
1/D. That is, halving D doubles the depth of focus. Now,
changing the f-number from 8 to 16 halves the diameter of the
aperture, so the new depth of focus is 0.8mm.

---

9. A camera has the range of f-numbers 2, 2.8, 4, 5.6, 8, 11, 16.
If the exposure time is the same for a series of exposures,
one with each f-number, what is the effect on the intensity at
the film plane?

$I \propto \dfrac{1}{(f/\#)^2}$, so in going from f/2 to f/16 I varies approximately
as follows:
$$\frac{1}{4}, \frac{1}{8}, \frac{1}{16}, \frac{1}{32}, \frac{1}{64}, \frac{1}{128}, \frac{1}{256} .$$

The stops are so chosen that changing from one to the next
changes the intensity by a factor of 2 or $\frac{1}{2}$, very nearly,
depending on whether one goes down or up the f-numbers.

---

10. Calculate the depth of field for a camera, focused for infinity,
having an aperture f/4 and focal length 50mm. Find the hyperfocal distance.

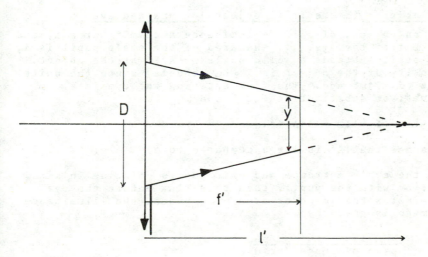

Fig. 17.10

The diameter of the aperture is $D = f'/(f/\#) = 50/4 = 12.5$mm. Hence, choosing the blur disc diameter to be $y = f'/1000 = 0.05$mm, an object at a distance ℓ will produce this size of blur disc when its image distance ℓ' is given by

$$\frac{y}{\ell'-f'} = \frac{D}{\ell'} \quad , \text{ from similar triangles in figure } 17.10.$$

Rearranging this for ℓ' gives $\ell' = \frac{Df'}{D-y}$.

The object distance ℓ is given by the thin lens formula

$$\frac{1}{\ell} = \frac{1}{\ell'} - \frac{1}{f'} = \frac{(D-y)}{Df'} - \frac{1}{f'} = \frac{(D-y)-D}{Df'} = -\frac{y}{Df'} \ .$$

So $\ell = -\dfrac{Df'}{y} = -\dfrac{12.5 \times 50}{0.05} = -12500$mm $= -12.5$m.

The depth of field is from infinity to 12.5m when the camera is focused for infinity. By definition this object distance is the hyperfocal distance.

---

11. When terrestrial objects are viewed through a telescope the images produced can never be brighter than when the same objects are viewed by the eye alone. However, when stars are viewed with the telescope there is an increase in the image brightness (illuminance). Explain.

Extended objects (terrestrial objects) - unaided eye

Consider an object, of uniform luminance B cd.m^{-2}, area α, and distance ℓ from the eye. If the area of the eye's pupil is A, then the pupil subtends a solid angle $\omega = A/\ell^2$ at the object. The intensity of the object is $I = B\alpha$ candela, since the units of $B\alpha$ are cd.m^{-2}m^2 = cd. The flux entering the eye is $F = I\omega$ which gives, on substituting for I and ω

$$F = I\omega = B\alpha A/\ell^2 .$$

(This assumes that ℓ^2 is large compared to α)

Assuming the eye's entrance and exit pupils coincide in size and position with the pupil, then this flux enters the eye and illuminates the image of area α'. Hence, the illuminance of the image is given by

$$\frac{\text{flux}}{\text{area of image}} = \frac{B\alpha A}{\ell^2 \alpha'} \ .$$

Extended objects - aided eye

Suppose now the area of the telescope objective is A_o and its
exit pupil is the same size as the eye-pupil and in the same
position as the latter. Then all the light entering the
objective enters the eye. We are always neglecting absorption
and reflection, of course. The flux entering the eye is now
$B\alpha A_o/\ell^2$, assuming the length of telescope is negligible, so ℓ
is as before. The illuminance of the image on the retina, the
area of which is now α'', is

$$\frac{B\alpha A_o}{\ell^2 \alpha''} \ .$$

But, if the magnifying power of the telescope is M, $\alpha'' = M^2\alpha'$,
so the illuminance of the retinal image is

$$\frac{B\alpha A_o}{\ell^2 \alpha''} = \frac{B\alpha A_o}{\ell^2 M^2 \alpha'} \ .$$

But the ratio of the objective's area to the area of the exit
pupil equals the square of the magnifying power of the telescope; that is, $M^2 = A_o/A$, where the eye pupil diameter is the
same as the exit pupil diameter.

Hence, the image's illuminance is

$$\frac{B\alpha A_o}{\ell^2 M^2 \alpha'} = \frac{B\alpha A}{\ell^2 \alpha'} \ ,$$

having put $A_o/M^2 = A$. But this is the same as with the unaided
eye, which is what we were asked to explain.

Note in passing that for the unaided eye

$$\frac{\alpha}{\alpha'} = \frac{1}{m^2} = \frac{(\ell/n)^2}{(\ell'/n')^2} = (n')^2\frac{\ell^2}{(\ell')^2} \ , \quad \text{since } n=1,$$

so the illuminance of the retinal image is

$$\frac{B\alpha A}{\ell^2 \alpha'} = \frac{BA}{\ell^2} \cdot (n')^2\frac{\ell^2}{(\ell')^2} = BA\left(\frac{n'}{\ell'}\right)^2 \ .$$

The term in brackets at the R.H.S. is constant for the eye, ℓ'
being the reduced eye's length and n' the eye's refractive index,
so the illuminance of the retinal image is proportional to
the object's luminance B and the pupil diameter A. It is most
important to note that the retinal image's illuminance does not
depend on the object's distance, ℓ.

Point objects - stars

The reason why there was no change in retinal image illuminance
with an extended object is as follows. When using the telescope the flux entering the eye was increased by $A_0/A = M^2$, but
the area of the retinal image was also increased by the same
factor $(\alpha''/\alpha' = M^2)$ so that the flux density (illuminance)
remained the same. With a point object, which a star is
effectively, the flux increases by M^2 when using the telescope
but the image size is still a point. Hence, the image appears
M^2 times brighter, neglecting losses.
